AF595075

Anatomical Variation and Clinical Diagnosis

Anatomical Variation and Clinical Diagnosis

Editor

Heather F. Smith

MDPI • Basel • Beijing • Wuhan • Barcelona • Belgrade • Manchester • Tokyo • Cluj • Tianjin

Editor
Heather F. Smith
Department of Anatomy
Midwestern University
Glendale
United States

Editorial Office
MDPI
St. Alban-Anlage 66
4052 Basel, Switzerland

This is a reprint of articles from the Special Issue published online in the open access journal *Diagnostics* (ISSN 2075-4418) (available at: www.mdpi.com/journal/diagnostics/special_issues/Anatomical_Variation).

For citation purposes, cite each article independently as indicated on the article page online and as indicated below:

LastName, A.A.; LastName, B.B.; LastName, C.C. Article Title. *Journal Name* **Year**, *Volume Number*, Page Range.

ISBN 978-3-0365-1334-8 (Hbk)
ISBN 978-3-0365-1333-1 (PDF)

Contents

About the Editor

Heather F. Smith

Dr. Heather F. Smith is a professor of anatomy at Midwestern University in Glendale, Arizona. She received her PhD from Arizona State University and completed her postdoctoral training at the University of Arizona's College of Medicine Phoenix. Her research focuses on anatomical variation, functional morphology, and paleontology. She teaches human gross anatomy to medical and graduate healthcare students.

Editorial

Anatomical Variation and Clinical Diagnosis

Heather F. Smith [1,2]

1 Department of Anatomy, College of Graduate Studies, Midwestern University, Glendale, AZ 85308, USA; hsmith@midwestern.edu; Tel.: +1-623-572-3726
2 School of Human Evolution and Social Change, Arizona State University, Tempe, AZ 85287, USA

Keywords: anatomy; anatomical variation; special issues

In the anatomical sciences, it has long been recognized that the human body displays a range of morphological patterns and arrangements, often termed "anatomical variation". Such variations are quite common and often have no noticeable impact on patient health. However, some variations may cause or contribute to significant medical conditions. An understanding of normal anatomical variation is vital for performing a broad range of surgical treatment modalities and other medical procedures. However, despite their importance for effective diagnosis and treatment, such variations are often overlooked in medical school curricula and clinical practice. Recent advances in imaging techniques and a renewed interest in variation in dissection-based gross anatomy laboratories have facilitated the identification of many such variants. This Special Issue of Diagnostics includes invited and contributed papers that highlight previously under-recognized anatomical variations and discuss them in a clinical context. In particular, the Issue focuses on variants that have specific implications for diagnosis and treatment and explores their potential consequences.

Several studies in this Special Issue present new information that has implications for surgical procedures, especially of the head and neck. Degaga and colleagues described variations in the pneumatization and septation of sphenoid sinus diagnosed using computed tomography (CT) scans, which have implications for trans-sphenoidal surgery [1]. Wang and colleagues evaluated differences in bone parameters at various dental implant sites in the mandible and maxilla [2]. Using cone-beam CT, they found that both cancellous bone density and cortical bone thickness vary by implant site, suggesting that these parameters should be evaluated prior to surgery. Thomas and colleagues presented new information regarding anatomical variations in the recurrent laryngeal nerve [3]. They presented a large-scale cadaveric evaluation and found that extra-laryngeal nerve branches are extremely common, suggesting the need to utilize left-sided approaches for surgical procedures of the neck whenever possible.

Other studies presented implications for surgical and other invasive interventions of the pelvis and extremities. Palomo-Lopez and colleagues documented the anatomic and histological features of the insertion of the extensor digitorum longus tendon on the proximal phalanx of the second toe [4]. They described for the first time the relationship between the tendon, distal nail matrix, and bone, which they explain should be used as reference landmarks during surgical procedures. Granite and colleagues presented a cadaveric study and review of the aberrant obturator artery, which is relevant to pelvic and groin surgeries [5]. Kostov and colleagues evaluated and described a series of anatomical variations in the ureters, pelvic vessels, and nerves that are relevant to the common surgical procedure of pelvic lymphadenectomy [6]. Finally, in a potentially revolutionary study, Garner, Plochocki, Hall, and colleagues describe how anatomical variation affects the potential for iatrogenic damage during episiotomy [7]. They discovered that episiotomies involving a midline incision resulted in a reduced risk of injury to nerves, erectile tissue, muscles and glands of the region than those involving mediolateral incisions. In the upper extremity, Sawyer and colleagues clarified the branching and innervation patterns

Citation: Smith, H.F. Anatomical Variation and Clinical Diagnosis. *Diagnostics* **2021**, *11*, 247. https://doi.org/10.3390/diagnostics11020247

Received: 27 January 2021
Accepted: 3 February 2021
Published: 5 February 2021

Publisher's Note: MDPI stays neutral with regard to jurisdictional claims in published maps and institutional affiliations.

of the radial nerve in the forearm [8]. They revealed significant variability in the posterior forearm, including occasional innervation of the brachialis and the superficial radial branch periodically innervating the brachioradialis.

A number of studies revealed anatomical and developmental bases for various pathologies and clinical presentations. Lee and colleagues described longitudinal changes in the tibiofibular relationship in patients with hereditary multiple exostoses [9]. A previously undescribed relationship was revealed in which the tibia grows relatively faster than the fibula, resulting in valgus angulation. Mirande and colleagues described two cases of retroesophageal aberrant right subclavian artery (ARSA) and demonstrated the enlarged size of the vessel relative to a normal sample [10]. As patients with this condition may present with dysphagia or be asymptomatic, they suggest that the relative size of the ARSA may dictate the severity of the associated symptomatology. Dyches and colleagues describe the acute angle of the celiac trunk in individuals with median arcuate ligament syndrome (MALS), thus explaining the mechanism of arterial compression in this condition [11]. They also described how, contrary to popular belief, the origin of the celiac trunk is not more superiorly positioned in patients with MALS compared to the general population. Grande-del-Arco and colleagues described the effect of tube angulation on the distortion of the metatarsal I head shape and clarified the correct shape of the metatarsal head in anatomical dissection [12].

Finally, Petto and colleagues presented a case of a rare anatomical variant, musculus sternalis [13]. They describe the unusual anatomy and explain how its finding in a gross anatomical teaching laboratory setting informed healthcare students about the importance of appreciating variation in diagnosis and treatment.

These papers present a variety of intriguing, clinically relevant anatomic variants, and explain their implications for surgical interventions, developmental understanding, and diagnoses. They emphasize the wealth of information that is still unknown about human anatomy. These studies also highlight the important contributions of body donors to medical education and the advancement of medical treatments, as many studies were conducted using cadaveric specimens. It is our sincere hope that the information in this Special Issue will be applied to advancing the diagnosis and treatment of patients with anatomic variants.

Conflicts of Interest: The author declare no conflict of interest.

References

1. Degaga, T.K.; Zenebe, A.M.; Wirtu, A.T.; Woldehawariat, T.D.; Dellie, S.T.; Gemechu, J.M. Anatomographic Variants of Sphenoid Sinus in Ethiopian Population. *Diagnostics* **2020**, *10*, 970. [CrossRef] [PubMed]
2. Wang, S.-H.; Shen, Y.-W.; Fuh, L.-J.; Chen, C.-F.; Tsai, M.-T.; Huang, H.-L.; Hsu, J.-T. Relationship between Cortical Bone Thickness and Cancellous Bone Density at Dental Implant Sites in the Jawbone. *Diagnostics* **2020**, *10*, 710. [CrossRef] [PubMed]
3. Thomas, A.M.; Fahim, D.K.; Gemechu, J.M. Anatomical Variations of the Recurrent Laryngeal Nerve and Implications for In-jury Prevention during Surgical Procedures of the Neck. *Diagnostics* **2020**, *10*, 670. [CrossRef] [PubMed]
4. Palomo-López, P.; Losa-Iglesias, M.E.; Becerro-De-Bengoa-Vallejo, R.; Rodríguez-Sanz, D.; Calvo-Lobo, C.; Murillo-González, J.; López-López, D. Anatomic and Histological Features of the Extensor Digitorum Longus Tendon Insertion in the Proximal Nail Matrix of the Second Toe. *Diagnostics* **2020**, *10*, 147. [CrossRef] [PubMed]
5. Granite, G.; Meshida, K.; Wind, G. Frequency and Clinical Review of the Aberrant Obturator Artery: A Cadaveric Study. *Diagnostics* **2020**, *10*, 546. [CrossRef] [PubMed]
6. Kostov, S.; Kornovski, Y.; Slavchev, S.; Ivanova, Y.; Dzhenkov, D.; Dimitrov, N.; Yordanov, A. Pelvic Lymphadenectomy in Gynecologic Oncology—Significance of Anatomical Variations. *Diagnostics* **2021**, *11*, 89. [CrossRef] [PubMed]
7. Garner, D.; Patel, A.; Hung, J.; Castro, M.; Segev, T.; Plochocki, J.; Hall, M. Midline and Mediolateral Episiotomy: Risk Assessment Based on Clinical Anatomy. *Diagnostics* **2021**, *11*, 221. [CrossRef]
8. Sawyer, F.K.; Stefanik, J.J.; Lufler, R.S. The Branching and Innervation Pattern of the Radial Nerve in the Forearm: Clarifying the Literature and Understanding Variations and Their Clinical Implications. *Diagnostics* **2020**, *10*, 366. [CrossRef] [PubMed]
9. Lee, J.H.; Rathod, C.M.; Park, H.; Lee, H.; Rhee, I.; Kim, H.W. Longitudinal Observation of Changes in the Ankle Alignment and Tibiofibular Relationships in Hereditary Multiple Exostoses. *Diagnostics* **2020**, *10*, 752. [CrossRef]
10. Mirande, M.H.; Durhman, M.R.; Smith, H.F. Anatomic Investigation of Two Cases of Aberrant Right Subclavian Artery Syn-drome, including the effects on external vascular dimensions. *Diagnostics* **2020**, *10*, 592. [CrossRef] [PubMed]

11. Dyches, R.P.; Eaton, K.J.; Smith, H.F. The Roles of Celiac Trunk Angle and Vertebral Origin in Median Arcuate Ligament Syn-drome. *Diagnostics* **2020**, *10*, 76. [CrossRef] [PubMed]
12. Grande-Del-Arco, J.; Becerro-De-Bengoa-Vallejo, R.; Palomo-López, P.; López-López, D.; Calvo-Lobo, C.; Pérez-Boal, E.; Losa-Iglesias, M.E.; Martin-Villa, C.; Rodríguez-Sanz, D. Radiographic Analysis on the Distortion of the Anatomy of First Metatarsal Head in Dorsoplantar Projection. *Diagnostics* **2020**, *10*, 552. [CrossRef] [PubMed]
13. Petto, A.J.; Zimmerman, D.E.; Johnson, E.K.; Gauthier, L.; Menor, J.T.; Wohkittel, N. Exploring Anatomic Variants to Enhance Anatomy Teaching: Musculus Sternalis. *Diagnostics* **2020**, *10*, 508. [CrossRef] [PubMed]

Article

The Roles of Celiac Trunk Angle and Vertebral Origin in Median Arcuate Ligament Syndrome

Ryan P. Dyches [1], Kelsey J. Eaton [1] and Heather F. Smith [2,3,*]

[1] Department of Osteopathic Manipulative Medicine, Arizona College of Osteopathic Medicine, Midwestern University, Glendale, AZ 85308, USA; rdyches65@midwestern.edu (R.P.D.); keaton15@midwestern.edu (K.J.E.)

[2] Department of Anatomy, Arizona College of Osteopathic Medicine, Midwestern University, Glendale, AZ 85308, USA

[3] School of Human Evolution and Social Change, Arizona State University, Tempe, AZ 85287, USA

* Correspondence: hsmith@midwestern.edu; Tel.: +1-623-572-3726

Received: 4 January 2020; Accepted: 29 January 2020; Published: 31 January 2020

Abstract: Median arcuate ligament syndrome (MALS) is a rarely diagnosed condition resulting from compression of the celiac trunk (CT) by the median arcuate ligament (MAL) of the diaphragm. Ischemia due to reduced blood flow through the CT and/or neuropathic pain resulting from celiac ganglion compression may result in a range of gastrointestinal symptoms, including nausea, postprandial discomfort, and weight loss. However, the mechanism of compression and its anatomical correlates have been incompletely delineated. It has been hypothesized that CT angle of origination may be more acute in individuals with MALS. Here, frequency of anatomical variation in the MAL and CT were assessed in 35 cadaveric subjects (17M/18F), including the vertebral level of origin of CT and superior mesenteric artery (SMA), the distance between CT and MAL and SMA, the angles of origination of CT and SMA, the diameter at the CT base, and MAL/CT overlap. Females exhibited significantly higher rates of inferred MAL/CT overlap than males. Significant correlations were revealed between MAL/CT overlap and angles of origination of the CT and SMA. Vertebral level of origin of the CT in individuals with MAL/CT overlap was not significantly more superior than in those without. This study also revealed a significant relationship between MAL/CT overlap and angle of origination of the CT, which has clinical implications for understanding the anatomy associated with MALS.

Keywords: median arcuate ligament syndrome; celiac artery; celiac artery compression syndrome; diaphragm; superior mesenteric artery

---

1. Introduction

Median arcuate ligament syndrome (MALS), also known as celiac artery compression syndrome or Dunbar syndrome, is a rarely diagnosed disease resulting from compression of the celiac trunk (CT) by the median arcuate ligament (MAL) [1] (Figure 1). The MAL is an arch-like fascial structure of the diaphragm linking the right and left diaphragmatic crura. While it has been noted as early as 1917 that the CT "is not infrequently partly covered by the diaphragm" [2], the precise relationship of the CT to the MAL was not extensively studied until recent years [3–7].

Embryologically, the abdominal diaphragm develops from its precursor, the septum transversum, which originates from mesenchyme in the mid-cervical region around embryonic day 22 [8]. Due to differential growth of the anterior and posterior regions of the embryo, the septum transversum appears to descend, and by week eight it reaches the level of the thoracic vertebrae [9]. The two crura (legs) of the diaphragm develop as muscular extensions that attach to the lateral sides of the lumbar vertebrae [9]. The crura are united in the midline by a tendinous band of fascia, the MAL. The aortic hiatus, an opening between the diaphragm and vertebral column around T12, permits passage of the abdominal aorta through the diaphragm [9]. Simultaneous to the development of the diaphragm, the CT, superior mesenteric artery (SMA), and inferior mesenteric artery (IMA) each form when the respective pair of developing segmental arteries converges at the midline of the abdominal aorta [8]. The CT develops to provide arterial supply to the foregut, the SMA to the midgut, and IMA to the hindgut. As the gut tube develops, the origins of these three unpaired visceral branches migrate caudally until they reach their final vertebral level around the end of month two [8]. The most common vertebral positions reported in adults are: T12 for the CT, L1 for SMA, and L3 for IMA [9]. Thus, the aortic hiatus, bounded by the MAL, frequently approximates the vertebral level of the CT, leaving little space between them. In some cases, the MAL may even overlap the CT, a condition which may result in impingement of the CT.

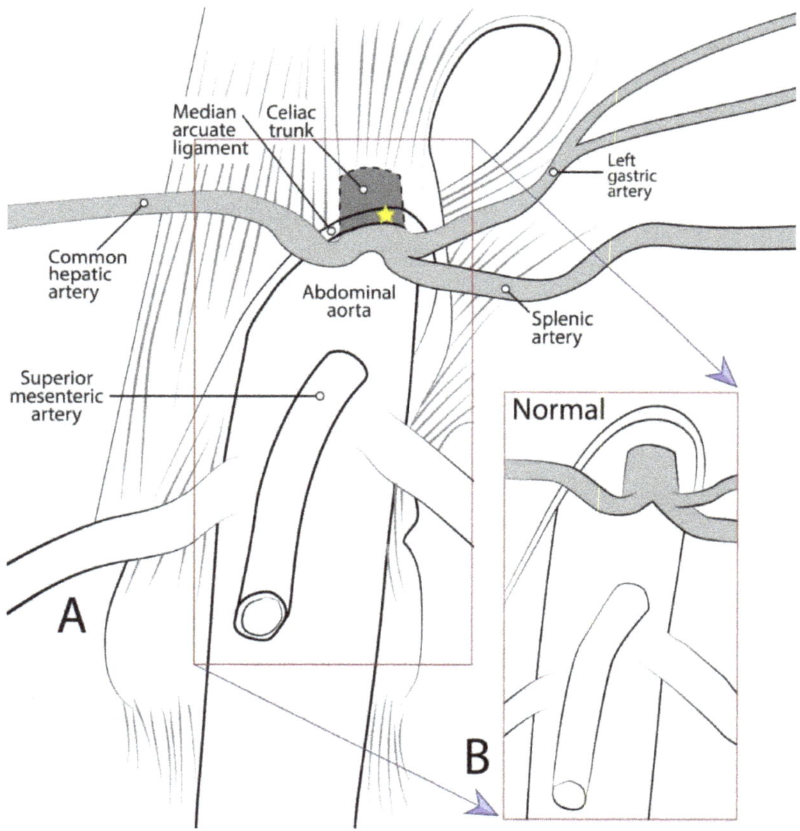

Figure 1. Diagrammatic representation of celiac trunk anatomy, showing: (**A**) The median arcuate ligament compressing the celiac trunk as in median arcuate ligament syndrome; and (**B**) normal anatomical condition.

The reported incidence of MALS is low; however, it has been postulated that in 10% to 24% of the population, the MAL crosses the aorta at an atypically inferior anatomic level and subsequently results in compression by the MAL [10]. It is hypothesized that MALS-related pain is both neuropathic and due to vascular compression in cause; namely, foregut ischemia pain results from decreased blood flow and chronic compression, as well as overstimulation of the celiac ganglia contributing to sympathetic neuropathic pain [1]. The celiac ganglia are collateral sympathetic ganglia that lie adjacent to the CT along its anterolateral sides [9]. They are the largest sympathetic ganglia in the body, and provide the sympathetic innervation to the foregut organs [9]. While they are typically paired, there can be up to five celiac ganglia, all of which are connected via a complex neural network, the celiac plexus, receiving preganglionic sympathetic fibers from T5–T12 [9]. Since the celiac ganglia are responsible for sympathetic innervation to, and visceral pain from, the foregut, impingement of the ganglia can result in a range of symptoms including radiating foregut discomfort, nausea, vomiting, epigastric fullness, and delayed gastric emptying [1]. Thus, the MALS symptom complex may include a range of expressions resulting from both vascular ischemia and/or neuropathic pain and overstimulation.

The vast majority of patients with partial CT compression are asymptomatic because collateral circulation typically prevents the development of symptoms [11]. If the compression is severe and symptomatic, however, a diagnosis of MALS is considered. The typical MALS patient presentation is a young, thin woman with symptoms of nausea, vomiting, early satiety, difficulty gaining weight, postprandial epigastric pain, and an abdominal bruit [12,13]. Further workup with Doppler ultrasound and/or computed tomographic (CT-scan) imaging is required for definitive diagnosis [14–17]. Indication for surgery is established in these otherwise healthy, symptomatic patients by showing a baseline celiac velocity >200 cm/m^2 and stenosis of the CA on CT-scan or MR angiography [12,13].

Although the superior mesenteric artery (SMA) is commonly described as originating 1.0 cm inferior to the CT, several studies have demonstrated the CT and SMA to be immediately adjacent in 22–58% of cases [3,4,18]. It is postulated that the close proximity of the two arteries may be evidence of CT compression by the MAL. In addition, one study found that up to 37% of cadavers had evidence of kinking of the CT [4]. A different study on fresh cadaveric specimens demonstrated that the CT origin was at or above the MAL in 33% of subjects [19]. However, it was unclear whether the CT originated more superiorly than normal, if the diaphragm extended further inferiorly, or if there was a combination of the two factors. A final possible indicator of CT compression is the CT angle of origination. It has been postulated that if there is an acute deviation of the left and right diaphragmatic crura, the MAL may cause constriction of the CT [20]. Despite these suggestions, the relationship between CT angle of origination and MAL compression has not been systematically assessed. This study aimed to increase the knowledge regarding the prevalence of MAL/CT overlap, the approximation and origin vertebral levels of the CT and the SMA, and the angle of origination of the CT, with the goal of better understanding the anatomical correlates of MALS. The study also sought to identify whether the vertebral level of the CT are associated with MAL/CT overlap.

2. Materials and Methods

2.1. Samples

Thirty-five cadaveric specimens (17M/18F) from the gross anatomy teaching laboratories at Midwestern University were assessed to determine the frequency of anatomical variation in the MAL and CT, especially MAL/CT overlap. Body donors ranged in age from 51–93 years old with a mean age of 77.4 years. Following student dissection, which included removal of the gastrointestinal organs, the integrity and presence of the MAL and CT were first evaluated. Additional dissection and preparation were necessary to further reveal the MAL and its associated neurovascular structures in several cadavers. Cadavers were embalmed with 10% formaldehyde and standard embalming protocols. Any specimens with gross gastrointestinal (GIT) abnormalities were removed from consideration. This study was determined to be IRB-exempt by the Midwestern University Institutional Review Board, due to the subjects being entirely cadaveric (#AZ 1259).

2.2. Data Collection

Data were then collected on several variables relating to size and relative positions of the structures of interest. First, the vertebral levels of the CT and SMA were assessed via manual palpation of the ribs and vertebrae. The CT was assessed for any evidence of overlap of the CT by the MAL or diaphragmatic crura by measuring the distance between the two structures. Negative distances were interpreted to indicate overlap. The diameter of the celiac trunk at its origin from the aorta and the distances between the CT and each of the MAL and SMA were measured using Mitutoyo sliding digital calipers. Finally, the angles of origination of the CT and SMA were measured using the U Protractor application on an iPhone.

2.3. *Statistical Analyses*

A series of statistical analyses were conducted to evaluate MAL/CT overlap and the other variables, and to determine whether sex differences existed in each variable. In particular, Chi-squared analyses were conducted to determine whether significant differences in frequency of MAL/CT overlap existed between the sexes. Analyses of Variances (ANOVA) were conducted to assess whether sex differences existed in any of the other variables, and to determine whether individuals with MAL/CT overlap exhibited significant differences in any of the other variables. Correlation analyses were performed to assess the relationship among each pair of variables. Since it is possible that some anatomical variables change with age, Partial Correlation analyses were also conducted controlling for age. All statistical analyses were performed using SPSS 25 (IBM Corp, Armonk, NY USA).

3. Results

3.1. *Comparison of Sexes and Rates of MAL/CT Overlap*

Approximately one-third of subjects (31.4%) were found to exhibit a morphology in which the diaphragm extended further inferiorly than the base of the CT, at least partially covering it (Figure 2; Table S1). Chi-squared tests revealed a highly significant difference in MAL/CT overlap between the sexes (Table 1; X2 = 12.39; $p < 0.001$), with females exhibiting a significantly higher frequency of MAL/CT overlap than males (44.4% vs. 21.4%). An ANOVA revealed significant differences in the origination angles of both CT ($F = 7.256$, $p = 0.011$) and SMA ($F = 15.084$, $p < 0.001$) between individuals with MAL/CT overlap and those without. However, there was no significant difference in the vertebral level of the CT or SMA between individuals with MAL/CT overlap versus those without. There was also no significant difference in diameter of CT or in the distance between the CT and SMA.

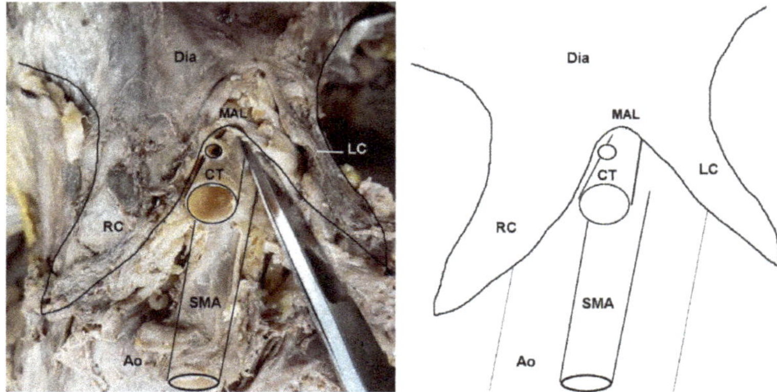

Figure 2. Dissection photo showing compression of the celiac trunk by the median arcuate ligament in a cadaveric subject: (**left**) Photo; (**right**) illustration. Abbreviations are as follows: Ao = aorta; CT = celiac trunk; Dia = diaphragm; LC = left crus of diaphragm; MAL = median arcuate ligament; RC = right crus of diaphragm; SMA = superior mesenteric artery.

Table 1. Cadaveric data on anatomical variables measured broken down by sex and MAL/CT overlap status.

	Origin CT	Origin SMA	Distance CT-MAL (mm)	Distance CT-SMA (mm)	Diameter CT (mm)	Angle Origination CT	Angle Origination SMA
Females	T11 = 16.7%, T12 = 50.0%, L1 = 33.3%	T12 = 38.9%, L1 = 38.9%, L2 = 22.2%	4.6	11.2	8.0	58.6°	58.4°
Males	T11 = 23.5%, T12 = 52.9%, L1 = 23.5%	T12 = 47.1%, L1 = 47.1%, L2 = 5.9%	8.7	10.0	8.9	72.1°	59.1°
MAL/CT overlap	T11 = 18.2%, T12 = 54.5%, L1 = 27.3%	T12 = 54.5%, L1 = 18.2%, L2 = 27.3%	11.1	10.6	8.7	72.4°	67.4°
MAL/CT non-overlap	T11 = 20.8%, T12 = 50.0, L1 = 29.2%	T12 = 37.5%, L1 = 54.2%, L2 = 8.3%	−3.3	10.7	7.8	49.4°	39.9°
Mean	**T11= 20%, T12= 51.4%, L1 = 28.6%**	**T12 = 42.9%, L1 = 14.3%, L2 = 42.9%**	**6.6**	**10.6**	**8.4**	**65.1°**	**58.8**

Abbreviations: CT = celiac trunk; MAL = median arcuate ligament; SMA = superior mesenteric artery. Mean values indicated in bold text.

3.2. Regression Results

The regression analyses revealed correlations between age and: external diameter of the celiac trunk ($r = 0.500$, $p = 0.002$), and angle of origination of the CT ($r = 0.467$, $p = 0.005$). Therefore, we also conducted a second set of analyses controlling for age through a partial correlation. When age was factored out, significant correlations were revealed between MAL/CT overlap and angles of origination of both the CT ($r = 0.466$, $p = 0.005$) and SMA ($r = 0.439$, $p = 0.009$), further supporting the results from the ANOVA (Table 2; Figure 3).

Table 2. Results of partial correlation analyses controlling for age among variables. Significant values are indicated in bold. It is interesting to note that the vertebral origin of the CT and SMA did not yield significant values.

	Diameter CT	Angle Origination CT	Angle Origination SMA	MAL/CT Overlap
Sex	0.247 ($p = 0.152$)	0.269 ($p = 0.118$)	0.15 ($p = 0.933$)	−0.289 ($p = 0.093$)
Vertebral origin CT	−0.022 ($p = 0.900$)	0.126 ($p = 0.472$)	−0.059 ($p = 0.738$)	0.005 ($p = 0.977$)
Vertebral origin SMA	−0.198 ($p = 0.254$)	−0.069 ($p = 0.693$)	0.082 ($p = 0.640$)	0.013 ($p = 0.943$)
Distance CT-MAL	**0.334 ($p = 0.050$)**	**0.360 ($p = 0.033$)**	0.170 ($p = 0.329$)	**−0.387 ($p = 0.022$)**
Distance CT-SMA	**−0.385 ($p = 0.022$)**	−0.201 ($p = 0.247$)	0.277 ($p = 0.107$)	0.010 ($p = 0.956$)
Diameter CT		**0.467 ($p = 0.005$)**	−0.009 ($p = 0.960$)	−0.231 ($p = 0.182$)
Angle origination CT			0.313 ($p = 0.067$)	**−0.425 ($p = 0.011$)**
Angle origination SMA				**−0.560 ($p < 0.001$)**

Abbreviations: CT = celiac trunk; MAL = median arcuate ligament; SMA = superior mesenteric artery.

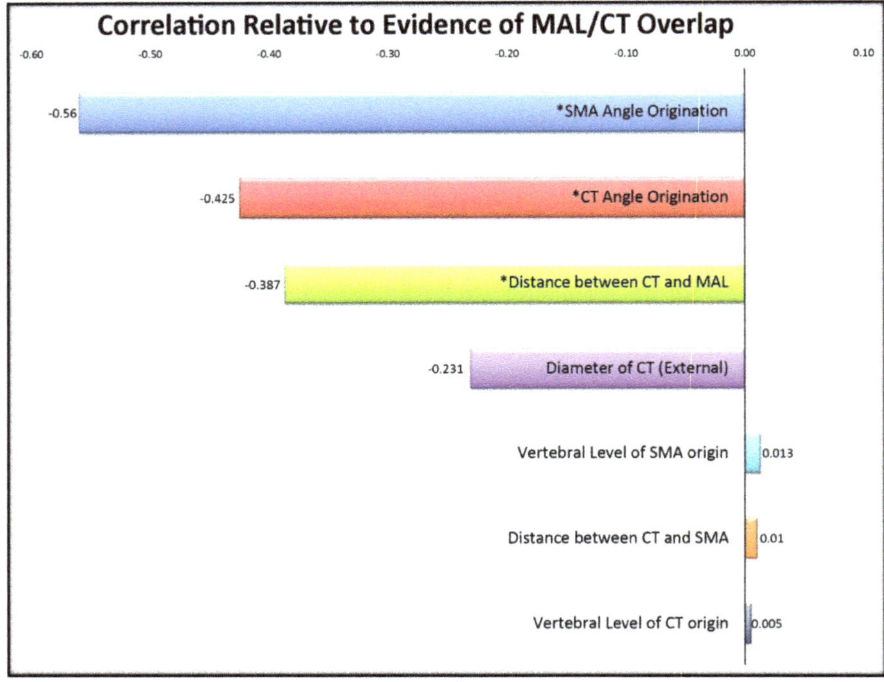

Figure 3. Correlations of cadaveric variables in relation to evidence of MAL/CT overlap. Significant values ($p < 0.05$) are denoted with an asterisk (*). Its significant negative correlation indicates that a smaller angle of origination suggests a greater likelihood of MAL/CT overlap.

3.3. Clinically Relevant Variants Identified

In addition to instances of MAL/CT overlap, several other clinically relevant anatomical variants of the CT were observed in the sample. In two female specimens, the CT arose deep to one of the diaphragmatic crura and coursed through its fibers to emerge into the abdominal cavity (Figure 4). These cases were tallied as instances of MAL/CT overlap; however, the mechanism of overlap was the crus rather than MAL. One male specimen displayed a rare condition of the left gastric artery branching directly off the abdominal aorta, rather than off the CT (Figure 5). In this case, the left gastric artery arose approximately 1 cm superiorly to the CT, and was covered by the MAL.

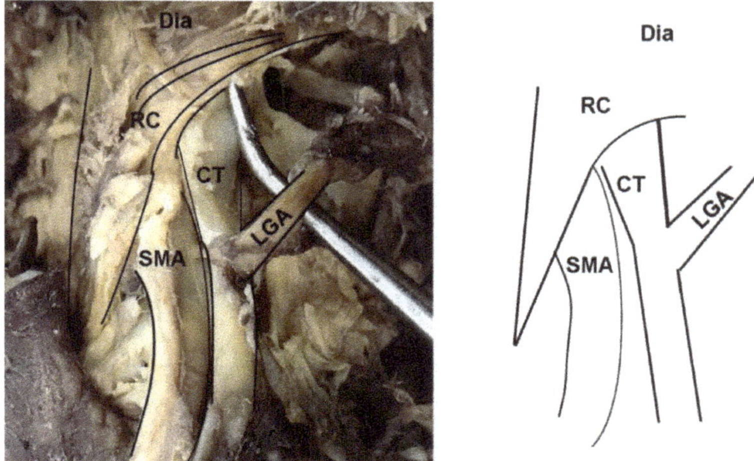

Figure 4. Dissection photo showing overlap of the celiac trunk by the right crus of the diaphragm: (**left**) Photo; (**right**) illustration. Abbreviations are as follows: CT = celiac trunk; Dia = diaphragm; LGA = left gastric artery; RC = right crus of diaphragm; SMA = superior mesenteric artery.

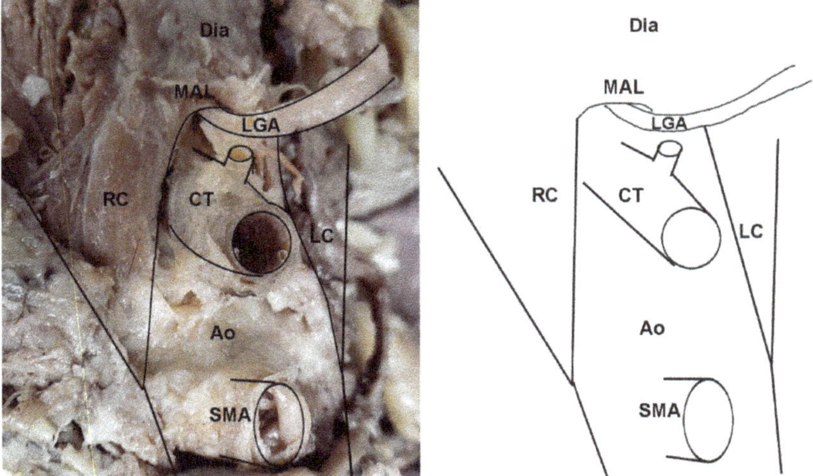

Figure 5. A notable variant displaying an uncommon branching of the left gastric artery (LGA) directly off the aorta in a cadaveric specimen: (**left**) Photo; (**right**) illustration. This example also demonstrates the direct overlap of the LGA by the median arcuate ligament. Abbreviations are as follows: Ao = aorta; CT = celiac trunk; Dia = diaphragm; LC = left crus of diaphragm; LGA = Left gastric artery; MAL = median arcuate ligament; RC = right crus of diaphragm; SMA = superior mesenteric artery.

4. Discussion

4.1. Angle of Celiac Trunk

The significant findings of this study, seen in Figure 3, demonstrate that when the CT angle of origin was more acute, there was an increased occurrence with MAL/CT overlap. Importantly, we believe this angle is the demonstration of the overlap, not the cause, in light of the correlation with the distance from the MAL implying the close proximity of the ligament. This confirms the source of overlap as being structurally induced by the ligament rather than in relation to an inherent angle.

4.2. Vertebral Origin of Celiac Trunk

Other studies have explored the branching height of the CT as it bifurcates from the abdominal aorta. It intuitively makes sense that with a more proximal branching point, the trunk and corresponding arteries will need to travel more acutely to exit from underneath the MAL. However, our findings did not support this hypothesis. As seen in Table 2, no correlation was found between the vertebral height of the CT and the observed MAL/CT overlap. This finding suggests that the mechanism of overlap may be the morphology of the MAL and diaphragm, rather than the height of the CT. In other words, individuals with MALS are more likely to have an inferiorly positioned or elongated diaphragm than a high CT. This interpretation is further supported by the two examples in which the CT was otherwise normally-positioned, but still emerged through the left crus of the diaphragm. Future studies comparing length of the crura between individuals with and without MAL/CT overlap could clarify this relationship.

4.3. Clinical Implications of MALS Findings

MALS significantly affects multiple health aspects of afflicted patients, creating both physical and emotional detriment. Patients undergo not only severe pain but the frustrations of symptomatology affecting activity participation, social interactions, as well as an effect on multiple body systems due to nutritional deficit. The diagnosis continues to be an elusive topic of understanding. Various established researchers and journals have ventured into determining its prevalence and clinical significance, yet its legitimacy is still debated. Previous literature has demonstrated high rates of variable CT compression by the MAL, in 10–50% of cases [19,21]. True symptomatology is minimal and intervention success remains imperfect, with only 50–80% of symptoms resolving after surgical release [22]. The deliberation over the validity of this syndrome lies within whether there is an undefined comorbid condition, as not all patients with compression visible on CT demonstrate symptoms [16]. Our results suggest that the angle of origination of the CT from the aorta may be useful as part of the diagnostic assessment of possible cases of MALS. This angle can be reliably determined using contrast-enhanced computed tomographic angiography (CTA). The goal is that with improved specificity of diagnostic measures, patients' conditions may be more rapidly treated and with greater confidence when using invasive measures.

Per prior reports, symptomatology is an important correlation when assessing compression. Similar to the phenomenon of magnetic resonance imaging (MRI) exposing vertebral disc abnormalities in >50% of asymptomatic individuals 30–39 years of age with disc degeneration [23], the MAL may apply compression without symptomology. An additional difficulty arises in that not all subjects express symptoms in a consistent manner. In a study by Harr et al. in 2014, young athletes were more likely to describe a stich feeling rather than the "classic" nausea, vomiting, and abdominal pain [24]. This subset of patients may already display a thin body habitus due to their athletic activities which may confound the classic presentation of MALS. Care providers should be aware of MALS in thin female athletes whose performance suddenly drops without otherwise evident causes.

4.4. Clinical Implications of Additional Anatomical Variants Identified

In addition to instances of full CT overlap by MAL, this study also revealed an alternate example of overlap of the foregut arterial supply by the diaphragm. In one specimen, the left gastric artery arose from the aorta superior to the CT and was covered by the MAL. While this condition would likely not be associated with the full range of symptoms as overlap of the entire CT, it could nonetheless result in reduced blood supply to the stomach, especially along the lesser curvature. This ischemia could result in similar, albeit less severe, symptoms of MALS, such as early satiety, postprandial discomfort, and/or nausea. However, in this condition, the celiac ganglion would likely be spared from overstimulation, and the blood flow to the liver, spleen, and pancreas would not be compromised.

4.5. Limitations

Our use of cadaveric specimens allows for assessment of arterial angles without diaphragmatic motion effect or muscular tension. While this is useful in assessing for pure angulation of the branching point, we recognize that this limits the assessment of the effects of breathing patterns on arterial compression, one that is predominantly found in exhalation. While a limitation, it is important to note that the ligament is the least mobile part of the diaphragm. Similarly, it is conceivable that dissecting these structures could alter their natural position in the body, or that the act of embalming the cadaver could alter the arterial angles. Further study of in vivo subjects, CT angiogram, and correlation with symptomatology is the next progression in this study, and will further elucidate whether CT angle is directly correlated with symptomatology. Finally, it should be acknowledged that the sample size is not large and is necessarily skewed in the direction of older aged individuals. However, within the sample, a broad range of stature, weight, and morphology was observed, suggesting that the sample captures as reasonably representative sample of older Americans.

5. Conclusions

This study revealed a high degree of variability in the relationship between the MAL and CT, identifying a high frequency of MAL/CT overlap. It also confirmed that MAL/CT overlap occurs significantly more often in women. While this study did not specifically address the reason for this sex difference, it may be related to an overall lower percentage of abdominal fat in women. Additionally, it revealed that MAL/CT overlap is not generally associated with a superiorly positioned CT. The angle correlation, a new perspective, allows another means of assessment in conjunction with the preexisting criteria of post stenotic dilation, flow variance, constriction. It is our belief that this information will allow for increased specificity in future studies of those with symptomatology, so as to aid in intervention decisions.

Supplementary Materials: The following are available online at http://www.mdpi.com/2075-4418/10/2/76/s1, Table S1: data file.

Author Contributions: Conceptualization, R.P.D. and H.F.S.; methodology, R.P.D. and H.F.S.; statistical analysis, H.F.S.; investigation, R.P.D., K.J.E., H.F.S.; resources, H.F.S.; data curation, R.P.D.; writing—original draft preparation, R.P.D.; writing—review and editing, R.P.D., K.J.E., H.F.S.; supervision, H.F.S.; project administration, H.F.S.; funding acquisition, H.F.S. All authors have read and agreed to the published version of the manuscript.

Funding: This research was funded by Midwestern University.

Acknowledgments: Funding for this research was provided by Midwestern University. The authors would like to thank the generous body donors whose cadavers formed the basis of this study. Thank you to Ashley Bergeron and the Anatomical Laboratories staff for their accommodation in the Anatomy laboratories. Figure 1 was created by Brent Adrian.

Conflicts of Interest: The authors declare no conflict of interest. The funders had no role in the design of the study; in the collection, analyses, or interpretation of data; in the writing of the manuscript, or in the decision to publish the results.

References

1. Kim, E.N.; Lamb, K.; Relles, D.; Moudgill, N.; DiMuzio, P.J.; Eisenberg, J.A. Median arcuate ligament syndrome—Review of this rare disease. *JAMA Surg.* **2016**, *151*, 471–477. [CrossRef]
2. Lipshutz, B. A composite study of the coeliac axis artery. *Ann. Surg.* **1917**, *65*, 159–169. [CrossRef]
3. Ali Mirjalili, S.; McFadden, S.L.; Buckenham, T.; Stringer, M.D. A reappraisal of adult abdominal surface anatomy. *Clin. Anat.* **2012**, *25*, 844–850. [CrossRef]
4. Katz-Summercorn, A.; Bridger, J. A cadaveric study of the anatomical variation of the origins of the celiac trunk and the superior mesenteric artery: A role in median arcuate ligament syndrome? *Clin. Anat.* **2013**, *26*, 971–974. [CrossRef]

5. Nasr, L.A.; Faraj, W.G.; Al-Kutoubi, A.; Hamady, M.; Khalifeh, M.; Hallal, A.; Halawani, H.M.; Wazen, J.; Haydar, A.A. Median arcuate ligament syndrome: A single-center experience with 23 patients. *Cardiovasc. Intervent Radiol.* **2017**, *40*, 664–670. [CrossRef]
6. Sugae, T.; Fujii, T.; Kodera, Y.; Kanzaki, A.; Yamamura, K.; Yamada, S.; Sugimoto, H.; Nomoto, S.; Takeda, S.; Nakao, A. Classification of the celiac axis stenosis owing to median arcuate ligament compression, based on severity of the stenosis with subsequent proposals for management during pancreatoduodenectomy. *Surgery* **2012**, *151*, 543–549. [CrossRef]
7. White, R.D.; Weir-McCall, J.R.; Sullivan, C.M.; Mustafa, S.A.R.; Yeap, P.M.; Budak, M.J.; Sudarshan, T.A.; Zealley, I.A. The celiac axis revisited: Anatomic variants, pathologic features, and implications for modern endovascular management. *Radiographics* **2015**, *35*, 879–898. [CrossRef]
8. Pansky, B. *Review of Medical Embryology*; Macmillan: New York, NY, USA, 1982; pp. 1–527.
9. Moore, K.L.; Dalley, A.F.; Moore, A. *Clinically Oriented Anatomy*, 8th ed.; Lippincott Williams & Wilkins: Philadelphia, PA, USA, 2017; pp. 1–1168.
10. Loukas, M.; Pinyard, J.; Vaid, S.; Kinsella, C.; Tariq, A.; Tubbs, R.S. Clinical anatomy of celiac artery compression syndrome: A review. *Clin. Anat.* **2007**, *20*, 612–617. [CrossRef]
11. Berney, T.; Pretre, R.; Chassot, G.; Morel, P. The role of revascularization in celiac occlusion and pancreatoduodenectomy. *Am. J. Surg.* **1998**, *176*, 352–356. [CrossRef]
12. Duncan, A.A. Median arcuate ligament syndrome. *Curr Treat. Options Cardiovasc. Med.* **2008**, *10*, 112–116. [CrossRef]
13. Jimenez, J.C.; Harlander-Locke, M.; Dutson, E.P. Open and laparoscopic treatment of median arcuate ligament syndrome. *J. Vasc Surg.* **2012**, *56*, 869–873. [CrossRef]
14. Aswani, Y.; Thakkar, H.; Anandpara, K.M. Imaging in median arcuate ligament syndrome. *BMJ Case Rep.* **2015**, bcr2014207856. [CrossRef]
15. Gruber, H.; Loizides, A.; Peer, S.; Gruber, I. Ultrasound of the median arcuate ligament syndrome: A new approach to diagnosis. *Med. Ultrason* **2012**, *14*, 5–9.
16. Horton, K.M.; Talamini, M.A.; Fishman, E.K. Median arcuate ligament syndrome: Evaluation with CT angiography. *Radiographics* **2005**, *25*, 1177–1182. [CrossRef]
17. Ozel, A.; Toksoy, G.; Ozdogan, O.; Mahmutoglu, A.S.; Karpat, Z. Ultrasonographic diagnosis of median arcuate ligament syndrome: A report of two cases. *Med. Ultrason.* **2012**, *14*, 154–157.
18. Paz, Z.; Rak, Y.; Rosen, A. Anatomical basis for celiac trunk and superior mesenteric artery entrapment. *Clin. Anat.* **1991**, *4*, 256–264. [CrossRef]
19. Lindner, H.H.; Kemprud, E. A clinicoanatomical study of the arcuate ligament of the diaphragm. *Arch. Surg.* **1971**, *103*, 600–605. [CrossRef]
20. Schweizer, P.; Berger, S.; Schweizer, M.; Schaefer, J.; Beck, O. Arcuate ligament vascular compression syndrome in infants and children. *J. Pediatr Surg.* **2005**, *40*, 1616–1622. [CrossRef]
21. Szilagyi, D.E.; Rian, R.L.; Elliott, J.P.; Smith, R.F. The celiac artery compression syndrome: Does it exist? *Surgery* **1972**, *72*, 849–863.
22. Kohn, G.P.; Bitar, R.S.; Farber, M.A.; Marston, W.A.; Overby, D.W.; Farrell, T.M. Treatment options and outcomes for celiac artery compression syndrome. *Surg. Innov.* **2011**, *18*, 338–343. [CrossRef]
23. Brinjikji, W.; Luetmer, P.H.; Comstock, B.; Bresnahan, B.W.; Chen, L.E.; Deyo, R.A.; Halabi, S.; Turner, J.A.; Avins, A.L.; James, K. Systematic literature review of imaging features of spinal degeneration in asymptomatic populations. *Am. J. Neuroradiol.* **2015**, *36*, 811–816. [CrossRef]
24. Harr, J.N.; Haskins, I.N.; Brody, F. Median arcuate ligament syndrome in athletes. *Surg. Endosc.* **2017**, *31*, 476. [CrossRef]

Article

Anatomic Investigation of Two Cases of Aberrant Right Subclavian Artery Syndrome, Including the Effects on External Vascular Dimensions

Mitchell H. Mirande [1,*], Madelyn R. Durhman [1] and Heather F. Smith [2,3]

1 Arizona College of Osteopathic Medicine, Midwestern University, Glendale, AZ 85308, USA; mdurhman96@midwestern.edu

2 Department of Anatomy, College of Graduate Studies, Midwestern University, Glendale, AZ 85308, USA; hsmith@midwestern.edu

3 School of Human Evolution and Social Change, Arizona State University, Tempe, AZ 85287, USA

* Correspondence: mmirande76@midwestern.edu; Tel.: +1-541-591-2117

Received: 27 June 2020; Accepted: 12 August 2020; Published: 14 August 2020

Abstract: The retroesophageal aberrant right subclavian artery (ARSA) is a variation of the aortic arch that occurs asymptomatically in most patients. However, when symptomatic, it is most commonly associated with dysphagia. ARSA has also been noted as a location of potentially severe aneurysms in some patients, as well as posing a risk during surgical interventions in the esophageal region. This case study analyzes two individuals with ARSA morphology in comparison to a normal sample in order to gain a better anatomical understanding of this anomaly, potentially leading to better risk assessment of ARSA patients going forward. The diameter of the ARSA vessel was found to be substantially larger than both the right subclavian artery and brachiocephalic trunk of the subjects with classic aortic arch anatomy. As many ARSA individuals are asymptomatic, we hypothesize that the relative size of the ARSA may dictate its contribution to the presence and/or severity of associated symptomatology.

Keywords: retroesophageal right subclavian artery; aberrant subclavian artery; arteria lusoria; dysphagia lusoria; kinked vertebral artery; artery tortuosity; Kommerell's diverticulum

---

1. Introduction

The retroesophageal aberrant right subclavian artery (ARSA) or arteria lusoria is a vascular morphology that has been periodically reported in the literature and represents the most common aortic arch anomaly [1,2]. The ARSA represents about 17% of aortic malformations [1,3], has an incidence of 0.2–2.5% [1,4,5], a prevalence of 0.7–2.0% [6] and has a female predominance [6–8]. Clinical presentations of the ARSA include dysphagia (difficulty swallowing), dyspnea (labored breathing), chest pain, cough and weight loss [5]. Case studies have been presented discussing the discovery of this anatomic variant while uncovering the root cause of an individual patient's symptoms, which if any, have been most commonly linked to dysphagia (difficulty swallowing) [1,2,7,8].

While ARSA often goes undiagnosed, certain imaging techniques can assist in its identification and diagnosis in patients with suspected symptoms. Barium contrast studies, computed tomography (CT) and magnetic resonance (MR) angiography have all demonstrated to be effective techniques for evaluating and diagnosing the ARSA condition [5,6,9–11]. In particular, CT angiograms, which carry the advantage of providing high spatial resolution imaging [6], can reveal the anomalous branching pattern and resulting vascular anomalies associated with ARSA [9]. Mahmodlou and colleagues described the importance of using CT to evaluate ARSA and other vascular anomalies prior to surgical interventions of the esophagus [5]. When an aberrant subclavian artery impinges on the esophagus causing dysphagia, barium contrast studies also may be a useful diagnostic tool [6,9,10].

The classic anatomic arrangement of the right subclavian artery is as a branch off the brachiocephalic trunk. However, in individuals with ARSA, the right subclavian artery branches directly off the posterior aortic arch (Figure 1) [1,3–5,7,8]. This morphology affects its position and course relative to the mediastinal structures. In particular, rather than coursing directly medially through the thorax, an ARSA typically takes a more posterior route, most often traveling posterior to the esophagus and trachea (Figure 1) [1–3]. This aberrant position may cause compression of the aforementioned structures, resulting in a range of clinical symptoms.

This case study analyzes a sample of cadavers including two cadaveric specimens with an ARSA from an empirical approach to help elucidate the anatomic development and clinical implications of the retroesophageal right subclavian variant.

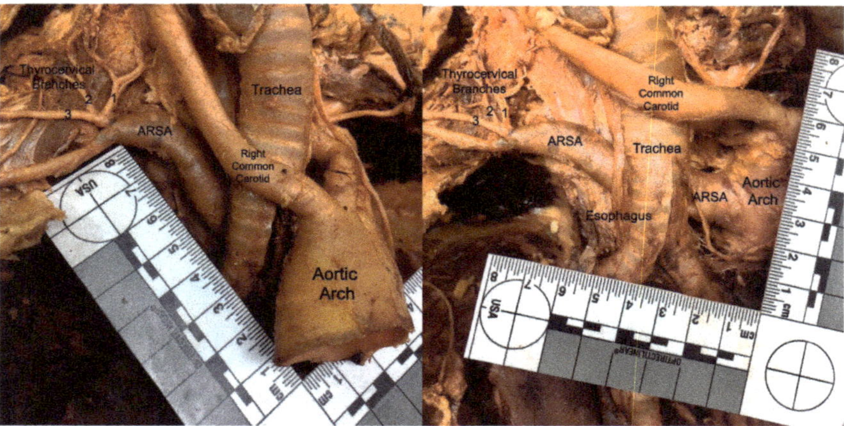

Figure 1. Dissection photo (left) showing the aberrant right subclavian artery (ARSA) of cadaver #1 (female), as it courses posterior to the esophagus and trachea. Dissection photo (right) shows the branching point of the aberrant right subclavian artery (ARSA), of cadaver #1 (female), from the posterior surface of the aortic arch. The aortic arch was retracted laterally to the left to gain this view.

2. Materials and Methods

2.1. Samples

The primary specimens in this case study were two cadavers with type G [3] retroesophageal right subclavian artery (ARSA) variants (1 M/1 F) from the anatomy teaching laboratory at Midwestern University. This unexpected finding in the teaching laboratory allowed for an opportunistic evaluation of the anatomy of this rare condition, including an analysis of external dimensions of the affected vasculature. In addition, twenty-three cadavers with classic aortic arch anatomy (11 M/12 F) were included in the study for comparative purposes. The causes of death were varied, and the subjects ranged in age from 63–95 at the time of death. Following student dissection of the cadavers, which consisted of removal of the lungs and heart, an initial assessment of the remaining aortic arch and branches was conducted. Further dissection was performed to open the visual field, allowing for arterial measurements to be collected. Cadavers with other abnormal aortic variants or loss of integrity of structures were excluded from consideration. This study was determined to be IRB-exempt by the Midwestern University Institutional Review Board (#AZ 1342).

2.2. Data Collection and Analysis

The primary goal of the present study was to evaluate the morphology of the two ARSA cadavers in comparison to the sample of non-ARSA cadavers. In particular, quantitative data on the external dimensions of vasculature that may have been impacted by ARSA were collected with Mitutoyo digital calipers. We investigated whether the right subclavian artery or its branches were larger in the ARSA cadavers. At each location, we measured the mediolateral width (a axis) and the anteroposterior width (b axis) at positions perpendicular to each other and calculated the surface area of each location using the equation for an ellipse ($A = \pi ab$, a = radius of a axis width, b = radius of b axis width). The dimensions of the following arteries were measured: base of aorta, base of brachiocephalic trunk (if present), bases and midpoints of left and right subclavian arteries, bases and midpoints of left and right common carotid arteries, bases of right and left vertebral arteries, bases of the thyrocervical trunk and each of its branches. The measurement for the base of the aorta was collected three centimeters from its first branch, measured on the anterior surface. We also measured eight distances to illustrate the branching distances of the aortic and proximal vessels: base of right subclavian artery to its first branch (usually vertebral artery), base of right subclavian artery to its midpoint, base of right subclavian artery to base of brachiocephalic trunk (normal cadavers only), base of right common carotid artery to base of brachiocephalic trunk (normal cadavers only), base of right subclavian artery to base of right common carotid artery (variants only), base of right subclavian artery to base of left common carotid artery (variants only), base of left subclavian artery to its first branch (usually vertebral artery) and base of left subclavian artery to its midpoint. These values were compared between the cadavers with classic anatomy versus the ARSA variants. Potential sex differences were evaluated statistically using Analyses of Variance in SPSS 25 (IBM Corp.). Following the completion of all the measurements, a structural assessment was conducted for each of the cadavers and the qualitative findings were also noted.

2.3. Ethical Standard Statement

Cadavers utilized in this study were obtained from Midwestern University Body Donation Program in Glendale, AZ, USA. The dissection of cadaveric specimens was performed according to The Common Rule regulations established in the Code of Federal Regulations (USA). The Institutional Review Board at Midwestern University indicated that IRB approval was not required for this project (#AZ 1342).

3. Results

3.1. Comparison of Variants vs. Normal Subjects

The ARSA variant was identified in two cadavers, one of which was a 63-year-old white male who died of a malignant neoplasm of the liver and the other was a white 73-year-old female who died of ischemic cardiomyopathy. Both ARSA cadavers exhibited a right subclavian artery that was substantially larger than both the right subclavian artery and the brachiocephalic trunk of their normal subject counterparts (Figure 2). The base of the ARSA at the posterior surface of the aortic arch was enlarged (variant avg = 394.17 mm^2 vs. normal avg. = 113.30 mm^2) and subsequently decreased in size throughout its route posterior to the esophagus. However, it consistently remained larger than the right subclavian artery of the normal cadavers along its route to the upper right extremity. Neither ARSA variant presented with a Kommerell's diverticulum (KOD) or had notable distal branching pattern anomalies. The bases of the aorta, right and left common carotid arteries and left subclavian artery were comparable in size between the two groups (Figure 2). Although there were no notable differences across the sample in the smaller downstream branches tested, the right thyrocervical trunk of the female ARSA cadaver #1 was the largest in the sample, measuring at about 10 mm larger than the mean of the non-ARSA cadavers (32.09 mm vs. 19.38 mm in non-ARSA).

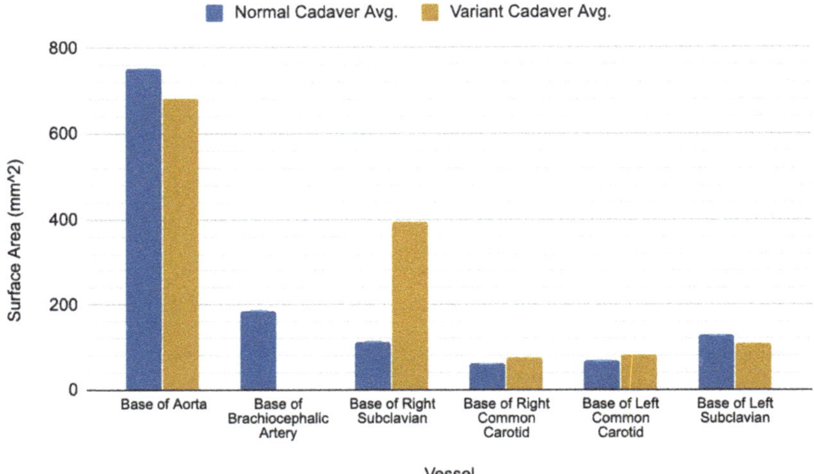

Figure 2. Comparative results showing the surface area measurements of aortic branches in the normal versus ARSA variant subjects. The surface areas at each location were averaged for each group.

3.2. Comparison of Sexes

Both female and male ARSA cadavers exhibited an enlarged right subclavian artery when compared to the average size of the normal cadavers; however, the disparity between the ARSA male and its normal counterparts was greater than the size disparity between the female variant and non-variant females (Supplementary Material Figure S1 and S2). In addition, the base of the aorta in the male variant was smaller compared to the average size of the normal cadavers, but the female variant base, although slightly larger, was more similar in size to the normal cadavers. In contrast, the female variant's left subclavian artery was smaller than the average normal females, but the male variant was similarly sized to its normal counterparts. The right and left common carotid arteries in both ARSA variants were similar in size to the normal specimens, but it was noted that the female variant showed slightly more divergence than the male variant from the average of the normal cadavers. Across the sample, males had a significantly larger right common carotid artery than females ($p = 0.041$), but there was no significant difference in size of the right subclavian artery ($p = 0.268$).

4. Discussion

4.1. Size and Clinical Implications of Retroesophageal Aberrant Right Subclavian Artery

The findings in this study demonstrate that the ARSA is significantly enlarged compared to a normal right subclavian and brachiocephalic trunk as it branches off the posterior aspect of the aortic arch (Table 1 and Supplementary Table S1). Importantly, its distinctly large size has significant clinical implications. One of the most common symptoms of this variant is dysphagia [1,7,8], often referred to as dysphagia lusoria, which is due to compression of the esophagus as the aberrant right subclavian artery travels past the esophagus posteriorly to the upper right extremity. However, in most cases, the condition is asymptomatic (90–93% of cases [8]) [5]. It is currently unclear why some patients present with symptoms of dysphagia, dyspnea and chest pain [5], while many others are asymptomatic. Additionally, it is unknown whether ARSA patients are at a higher risk of more severe complications such as aneurysm [12]. We believe that the relative size of the ARSA may help dictate its contribution to the presence and/or severity of associated symptomatology, as discussed in further detail below.

Table 1. Cadaveric data comparing the surface area (SA) of a variety of vessels in both normal and variant subjects of both sexes.

Vessel	Normal Male Cadaver Avg. SA (mm^2)	Variant Male Cadaver SA (mm^2)	Normal Female Cadaver Avg. SA (mm^2)	Variant Female Cadaver SA (mm^2)
Base of aorta	829.80	642.68	675.99	721.30
Base of brachiocephalic artery	198.70	–	175.41	–
Base of right subclavian artery	122.08	482.69	104.52	305.65
Base of right common carotid	67.33	71.78	53.64	80.87
Base of right vertebral artery	25.39	15.45	19.40	–
Base of left subclavian artery	143.98	142.32	111.33	78.17
Base of left common carotid artery	72.87	84.48	62.84	79.61
Base of left vertebral artery	22.97	21.83	22.36	16.76

Symptoms of the ARSA anomaly have been documented as presenting during the two extreme temporal periods of life due to anatomic changes [8,9]. In infants, the developing trachea is more flexible, which may allow the ARSA to compress the airway more easily, resulting in symptoms of stridor (wheezing caused by an obstructed airway) and respiratory tract infections [2,6,8]. A study conducted in 1993 found that 86% of infant patients with the ARSA morphology presented with stridor or recurrent respiratory infections [8,13]. On the other end of the temporal spectrum, in adults, respiratory symptoms co-occur with ARSA less frequently, as the adult trachea is more rigid. Instead, an inflexible trachea can lead to easier compression of the esophagus, which may contribute to the presentation of dysphagia as a primary symptom [8,13,14]. Other explanations of the presentation of dysphagia in older adults may result from the formation of an aneurysm, elongation of the aorta and increased rigidity of the esophagus or the vessel wall [3,8,15].

4.2. Coexisting Malformations of the Aberrant Right Subclavian Artery

An early study on ARSA morphology questioned whether the ARSA anomaly alone is sufficient to cause symptoms of dysphagia. Instead, it was argued that the presence of a coexisting bicarotid trunk (BCT), observed in 29% of their cases (85 out of 295 cases), was the cause for symptom presentation [3,8,16]. Subsequent studies have confirmed that the two conditions may co-occur [1,3,9,17]. However, a more recent study showed that while 71.2% of their 141 ARSA cases reported dysphagia as a symptom, only 19.2% (27/141) possessed the bicarotid morphology [8]. In addition, in many cases of a BCT co-occurring with an ARSA, barium-contrast examination indicates that the esophagus is compressed posteriorly and obliquely at the level of the aorta [3,4,6,9,15,18–20], suggesting that the ARSA is the primary contributor to esophageal compression and resulting dysphagia. There are also documented cases in which a BCT is absent or extremely mild, but dysphagia still presents [19,21]. These observations suggest that symptomology linked to the ARSA morphology is complex and multifactorial, with a variety of risk factors playing a role in the presentation of dysphagia or its variety of symptoms [6]. We propose that the larger that the diameter of the ARSA is, as it travels along its abnormal route posterior to the esophagus and trachea, the more likely it is to contribute to resulting compression and symptoms. Furthermore, the expansion of the lumen of the ARSA may result in compromise to the structural integrity of the vessel walls making aneurysms in this location more likely, causing more severe complications in these patients.

In addition to BCT, there are a variety of other vascular malformations that may coexist with the ARSA. It has been reported that the ARSA presentation may occur in Edwards (55%), Down's (7.9–29.6% in fetuses and 1.6–35.7% in adults), Turner (43%), Patau (50%), DiGeorge (14%), Potter, post-rubella and Noonan syndromes [7]. Outside of these syndromes, the ARSA may coexist with other cardiovascular malformations such as the tetralogy of Fallot (12%) [1] and can increase the likelihood of co-occurring visceral anomalies such as asplenia, gall bladder agenesis, esophageal atresia, trachea–esophageal fistula, anal atresia, lung lobation abnormalities, double uterus and vagina, renal anomalies and sacral spina bifida [7].

4.3. Aneurysm of the Aberrant Right Subclavian Artery

Aneurysms associated with the ARSA anomaly, although rare [12,22], are a severe complication when present. Understanding the risk that an ARSA patient has of developing an aneurysm would help clinicians better weigh the treatment options when caring for an ARSA patient. From our study, we determined that the ARSA has a drastically larger diameter at its branching point from the aorta than the corresponding arteries in normal individuals. We hypothesize that this enlarged vessel is at a higher risk of aneurysm formation due to a loss of vessel wall integrity during development. Additionally, this risk may increase with the presence of a Kommerell's diverticulum (KOD), which are more likely to develop in ARSA patients, due to aneurysmal degeneration [10]. The KOD, which is a diverticulum at the proximal descending aorta that gives rise to an aberrant subclavian artery in both the left and right aortic arch configurations, is a rare anomaly [23], which has been reported to occur in about 20–60% of individuals with an ARSA morphology [6]. If a Kommerell's diverticulum progresses to an aneurysm, it can carry a higher risk of mortality [2]. The true rupture or dissection rate of an aneurysm associated with a KOD is quite variable ranging from 0–50% [6]. Although more specific studies have reported first encounter rupture rates of 6% (2/33 cases) [24] and 19% (6/32 cases) [2,6] as well as dissection and/or rupture rates of 44% [24] or a sole dissection rate of 11% [6].

The subclavian artery is also characterized as a muscular type artery, which has a low percentage of elastic fibers in the tunica media [7], which may further explain the risk of aneurysm near the subclavian artery's branch point in ARSA patients [7]. Additionally, a study conducted by Kim et al. reported histologically the presence of cystic medial degeneration, which is the most common finding in a surgically resected specimen from patients with an aortic aneurysm or dissection, in the vascular wall of patients with a KOD [Kim]. The patients in this study had ARSA vessels which ranged in size from 6–10 mm and were accompanied by KODs which ranged from 15–45 mm [23]. The two cadaveric specimens in our study presented with ARSA vessels with a size of 19.18 mm (F) and 25.26 mm (M) with no associated KODs. These findings show that the presence of a KOD is a risk factor for an aneurysm or dissection to present. However, there are also large ARSA vessels that develop in the absence of a KOD which may not present with the same aneurysmal risk but result in a set of risks and symptoms due to their distinct morphology.

4.4. Embryological Origin of the Aberrant Right Subclavian Artery

During development, the right subclavian artery (RSA) derives from two sources. The proximal RSA develops from the right fourth aortic arch, whereas the distal portion of the RSA develops from the right seventh intersegmental artery on the dorsal aorta [2,7,25]. The ARSA variant morphology develops when the right fourth aortic arch fails to develop, which results in the right seventh intersegmental artery remaining attached to the dorsal aorta, which derives the ARSA formation [2,7,8]. In approximately 80% of cases, the ARSA takes a retroesophageal course, with an interesophageotracheal course occurring in 16.7% of cases and 5% of cases traveling anterior to the trachea [7,18].

4.5. Diagnosis for the Aberrant Right Subclavian Artery

Discovering a patient's ARSA anomaly is usually achieved incidentally [2] while investigating symptoms that may be comorbid among gastroesophageal reflux disease (GERD), respiratory disease and the ARSA variant. The use of barium contrast studies, computed tomography (CT) and magnetic resonance (MR) angiography have shown to be useful techniques for evaluating and diagnosing the ARSA anomaly.

4.5.1. Barium Contrast Study

The use of a barium contrast study is an effective way to identify posterior compression of the esophagus by the ARSA vessel [9–11]. Multiple studies have described the incidental discovery of the ARSA vessel while performing a barium contrast study on patients presenting with dysphagia [6,9,10].

This method can help identify the location of posterior compression as well as determine the pulsation of the ARSA on the esophagus [9]. Typically, the oblique view of the barium study will reveal a diagonal indentation in the posterior surface of the esophagus [6]. Although the barium study can identify location and relative size of the ARSA, it cannot be used to determine the dimensions of the ARSA or its relationship to surrounding organs [6]. Further diagnostic methods should be used to perform a risk assessment and to develop a treatment plan for the ARSA variant once identified [9,10].

4.5.2. Computed Tomography

Computed tomography (CT) is a useful tool to use in diagnosing the presence of an ARSA, although it has its disadvantages such as the use of iodine contrast media and irradiation [6,7]. The advantages of CT include its short scanning time, high spatial resolution imaging and availability, which can help in identifying the dimensions of the vessel, including the presence of a KOD [6,9–11]. Additionally, the use of a multidetector computed tomography (MDCT) scanner, with its three-dimensional images, can assist in the understanding of the patient's unique anatomy and assist providers during decisions regarding treatment strategy [6]. For example, in a case report by Mahmodlou, the use of CT revealed an ARSA vessel in a patient prior to a procedure on an esophageal cancer, which allowed surgeons to alter their surgical procedure to prevent potentially life-threatening injury to the ARSA [5].

4.5.3. Magnetic Resonance Imaging and Angiography

Magnetic resonance imaging (MRI) and magnetic resonance angiography (MRA), along with CT, are one of the current best diagnostic tools to use to identify the ARSA variant and its associated vascular anomalies such as a KOD [6,7,11]. MRI allows for visualization of any imaging plane and can identify the presence of esophageal compression and/or pulsation from the ARSA [6]. However, the use of MRI comes with a key disadvantage of requiring a greater length of time, up to 20–40 min for an examination [6]. The ARSA anomaly may also present with difficulty during MRA for angiographists or interventional cardiologists who utilize the brachial or radial approaches to reach the ascending aorta [1].

Although our cadaveric study does not contain imaging of these diagnostic techniques, various studies have shown the efficacy of these methods in identifying the ARSA morphology in living patients [5,9–11]. It is important for clinicians to be aware of the ARSA variant when using these diagnostic tools in relation to symptoms of dysphagia, dyspnea and chest pain [5,6].

4.6. *Treatment for the Aberrant Right Subclavian Artery*

Many patients with the ARSA morphology are asymptomatic and clinical intervention is not required [5]. When necessary, both medical and surgical treatments were described to treat the associated dysphagia, depending on the severity of the symptoms, variant anatomy and treatment response [18].

4.6.1. Surgical Treatments

Patients presenting with symptoms related to compression of the esophagus or trachea, and those with aneurysmal dilation or a KOD may require surgical intervention [1,4–6]. A key issue in deciding on treatment type depends on whether an accompanying aneurysm is present [10]. If the ARSA is non-aneurysmal, then the goal of surgical treatment is the closure of the ARSA at its origin and the revascularization of the right subclavian artery via a transposition of the ARSA to the right common carotid artery [4,10].

Another issue that influences the decision to proceed with surgical treatment is whether or not a KOD is present. Indications for surgical treatment of a KOD in an asymptomatic patient remain disputed [6]. However, it has been stated that the decision of treatment strategy for a KOD should be based on the available surgical expertise and the anatomy and comorbidities of the patient [6].

A recent study (2020) described a less-invasive approach using a left thoracoscopic procedure and a right supraclavicular incision, which facilitates a safer closure of the ARSA and avoids dramatic displacement of the esophagus and trachea [4]. This approach also permits an ideal degree of exposure of the relevant neurovasculature as well as reducing postoperative morbidity [4].

If the ARSA is associated with an aneurysm, then surgical treatment will be required to address the aneurysm prior to the transposition of the ARSA [10], due to the high rate of rupture (22.6% [3,26]). The indication for surgical intervention of an aneurysm or an associated KOD has been described based on size by many investigators, although the details of the measurement location and method has not been well reported or uniform [6]. A study conducted by Cina et al., in 2004 suggested that a diameter of 3 cm at the level of diverticulum orifice was the threshold for surgery in low-risk patients [6]. Additionally, a study conducted in 2014 by Vucemilo et al., stated that treatment is recommended if an ARSA patient has an aneurysm >3 cm in diameter, has a symptomatic or ruptured aneurysm or is symptomatic [2]. Due to the anatomical position of ARSA aneurysms, it is very difficult to treat them completely using only a single surgical approach (supraclavicular approach, median sternotomy or left thoracotomy) [22]. The use of left thoracotomy, median sternotomy or bilateral carotid—subclavian bypasses followed by a thoracic aortic endograft have been indicated for patients with the presence of an ARSA aneurysm [4]. In patients with a symptomatic ARSA, with or without an associated KOD, a systematic review conducted by Vucemilo et al. showed that the hybrid open and endovascular surgical approach is a safe and effective method of treatment, reporting resolution of thrombosis and presenting symptoms while decreasing aneurysmal sac size, length of hospital stay and complications [2].

4.6.2. Nonsurgical Treatments

For patients with less severe symptoms or those not wishing to move forward with surgical interventions, conservative management can be utilized which typically involves dietary modification as well as the use of proton pump inhibitors and prokinetic agents [11]. In a case series performed on six patients with dysphagia lusoria, conservative treatment with a proton pump inhibitor and a prokinetic agent showed to be 50% effective in treating symptoms [18].

4.7. Limitations

We concede that our analysis of only two ARSA specimens in this study limits our ability to make definitive claims about the significance of this morphology. However, compiling these data could assist future researchers in subsequent analyses and contribute to the understanding of this aortic anomaly. It is also possible that the embalming process or act of dissecting the cadavers may have altered the original positions of the structures studied; however, we consider preservation unlikely to have had dramatic effects on the measured dimensions. Further studies performed on in vivo subjects using CT angiograms would be the next step in furthering our functional comprehension of this variant.

5. Conclusions

This case study analyzed a rare aortic arch anomaly and demonstrated that not only the positioning of the branch point of the right subclavian artery deviated from the normal anatomic location, but that the size of the vessel was greatly enlarged compared to normal size. This observation is the beginning of gaining a better understanding of this anomaly and its range of presentations and comorbidities. This information and data set will add to prior studies and facilitate further research to gain a better understanding of the relationships between the ARSA variant and its associated variants and the potential risks these patients may face.

Supplementary Materials: The following are available online at http://www.mdpi.com/2075-4418/10/8/592/s1, Figure S1: Normal female vs. Variant female; Figure S2: Normal male vs. Variant male; Table S1: Data file.

Author Contributions: Conceptualization, M.H.M., M.R.D. and H.F.S.; data curation, M.H.M., M.R.D. and H.F.S.; formal analysis, M.H.M., M.R.D. and H.F.S.; investigation, M.H.M., M.R.D. and H.F.S.; methodology, M.H.M., M.R.D. and H.F.S.; project administration, H.F.S.; resources, H.F.S.; software, H.F.S.; supervision, H.F.S.; validation, H.F.S.; visualization, M.H.M. and H.F.S.; writing original draft, M.H.M., M.R.D. and H.F.S.; writing review & editing, M.H.M. and H.F.S. All authors have read and agreed to the published version of the manuscript.

Funding: This research received no funding.

Acknowledgments: The authors would like to thank the generous body donors whose cadavers formed the basis of this study. We thank Linda Walters for sharing her insight into this interesting variant.

Conflicts of Interest: The authors declare no conflict of interest. The funders of this study had no role in the design of the study; in the collection, analyses or interpretation of data; in the writing of the manuscript or in the decision to publish the results.

Abbreviations

ARSA	Aberrant right subclavian artery
RSA	right subclavian artery
BCT	bicarotid trunk
GERD	gastroesophageal reflux disease
KOD	Kommerell's diverticulum
CT	Computed Tomography
MRI	Magnetic Resonance Imaging
MRA	Magnetic Resonance Angiography
MDCT	multidetector computed tomography

References

1. de Araújo, G.; Junqueira Bizzi, J.W.; Muller, J.; Cavazzola, L.T. Dysphagia lusoria—Right subclavian retroesophageal artery causing intermittent esophageal compression and eventual dysphagia—A case report and literature review. *Int. J. Surg. Case Rep.* **2015**, *10*, 32–34. [CrossRef] [PubMed]
2. Vucemilo, I.; Harlock, J.A.; Qadura, M.; Guirgis, M.; Gowing, R.N.; Tittley, J.G. Hybrid repair of symptomatic aberrant right subclavian artery and Kommerell's diverticulum. *Ann. Vasc. Surg.* **2014**, *28*, 411–420. [CrossRef] [PubMed]
3. Rogers, A.D.; Nel, M.; Eloff, E.P.; Naidoo, N.G. Dysphagia lusoria: A case of an aberrant right subclavian artery and a bicarotid trunk. *ISRN Surg.* **2011**, *2011*, 819295. [CrossRef] [PubMed]
4. Amore, D.; Casazza, D.; Casalino, A.; Valente, T.; De Rosa, R.C.; Sangiuolo, P.; Curcio, C. Symptomatic Aberrant Right Subclavian Artery: Advantages of a Less Invasive Surgical Approach. *Ann. Thorac. Cardiovasc. Surg.* **2020**, *26*, 104–107. [CrossRef]
5. Mahmodlou, R.; Sepehrvand, N.; Hatami, S. Aberrant Right Subclavian Artery: A Life-threatening Anomaly that should be considered during Esophagectomy. *J. Surg Tech. Case Rep.* **2014**, *6*, 61–63. [CrossRef]
6. Tanaka, A.; Milner, R.; Ota, T. Kommerell's diverticulum in the current era: A comprehensive review. *Gen. Thorac. Cardiovasc. Surg.* **2015**, *63*, 245–259. [CrossRef]
7. Natsis, K.; Didagelos, M.; Gkiouliava, A.; Lazaridis, N.; Vyzas, V.; Piagkou, M. The aberrant right subclavian artery: Cadaveric study and literature review. *Surg. Radiol. Anat. SRA* **2017**, *39*, 559–565. [CrossRef]
8. Polguj, M.; Chrzanowski, Ł.; Kasprzak, J.D.; Stefańczyk, L.; Topol, M.; Majos, A. The aberrant right subclavian artery (arteria lusoria): The morphological and clinical aspects of one of the most important variations–a systematic study of 141 reports. *Sciences* **2014**, *2014*, 292734. [CrossRef]
9. Barone, C.; Carucci, N.S.; Romano, C. A Rare Case of Esophageal Dysphagia in Children: Aberrant Right Subclavian Artery. *Case Rep. Pediatr.* **2016**, *2016*, 2539374. [CrossRef]
10. Naqvi, S.E.H.; Beg, M.H.; Thingam, S.K.S.; Ali, E. Aberrant right subclavian artery presenting as tracheoesophageal fistula in a 50-year-old lady: Case report of a rare presentation of a common arch anomaly. *Ann. Pediatr Cardiol.* **2017**, *10*, 190–193. [CrossRef]
11. Reynolds, I.; McGarry, J.; Mullett, H. Aberrant right retroesophageal subclavian artery causing esophageal compression. *Clin. Case Rep.* **2015**, *3*, 897–898. [CrossRef] [PubMed]
12. Godlewski, J.; Widawski, T.; Michalak, M.; Kmieć, Z. Aneurysm of the aberrant right subclavian artery—A case report. *Pol. J. Radiol.* **2010**, *75*, 47–50. [PubMed]

13. van Son, J.A.; Julsrud, P.R.; Hagler, D.J.; Sim, E.K.W.; Pairolero, P.C.; Puga, F.J.; Schaff, H.V.; Danielson, G.K. Surgical treatment of vascular rings: The Mayo Clinic experience. *Mayo Clin. Proc.* **1993**, *68*, 1056–1063. [CrossRef]
14. McNally, P.R.; Rak, K.M. Dysphagia lusoria caused by persistent right aortic arch with aberrant left subclavian artery and diverticulum of Kommerell. *Dig. Dis. Sci.* **1992**, *37*, 1441–1449. [CrossRef]
15. Ulger, Z.; Ozyurek, A.R.; Levent, E.; Gurses, D.; Parlar, A. Arteria lusoria as a cause of dysphagia. *Acta Cardiol.* **2004**, *59*, 445–447. [CrossRef]
16. Klinkhamer, A.C. A berrant right subclavian artery. Clinical and roentgenology aspects. *Am. J. Roentgenol. Radium. Nucl. Med.* **1966**, *97*, 438–446. [CrossRef]
17. Kent, D.; Poterucha, H. Aberrant right subclavian artery and dysphagia lusoria. *N. Engl. J. Med.* **2002**, *346*, 1637. [CrossRef]
18. Epperson, M.V.; Howell, R. Dysphagia lusoria. *Curr. Opin. Otolaryngol. Head Neck Surg.* **2019**, *27*, 448–452. [CrossRef]
19. Karlson, K.J.; Heiss, F.W.; Ellis Jr, F.H. Adult dysphagia lusoria: Treatment by arterial division and reestablishment of vascular continuity. *Chest* **1985**, *87*, 684–686. [CrossRef]
20. Tong, E.; Rizvi, T.; Hagspiel, K.D. Complex aortic arch anomaly: Right aortic arch with aberrant left subclavian artery, fenestrated proximal right and duplicated proximal left vertebral arteries-CT angiography findings and review of the literature. *Neuroradiol. J.* **2015**, *28*, 396–403. [CrossRef]
21. Carrizo, G.J.; Marjani, M.A. Dysphagia lusoria caused by an aberrant right subclavian artery. *Tex. Heart Inst. J.* **2004**, *31*, 168–171. [PubMed]
22. Kamiya, H.; Knobloch, K.; Lotz, J.; Bog, A.; Lichtenberg, A.; Hagl, C.; Kallenbach, K.; Haverich, A.; Krack, M. Surgical treatment of aberrant right subclavian artery (arteria lusoria) aneurysm using three different methods. *Ann. Thorac. Surg.* **2006**, *82*, 187–190. [CrossRef]
23. Kim, K.M.; Cambria, R.P.; Isselbacher, E.M.; Baker, J.N.; LaMuraglia, G.M.; Stone, J.R.; MacGilivray, T.E. Contemporary surgical approaches and outcomes in adults with Kommerell diverticulum. *Ann. Thorac Surg.* **2014**, *98*, 1347–1354. [CrossRef] [PubMed]
24. Cinà, C.S.; Althani, H.; Pasenau, J.; Abouzahr, L. Kommerell's diverticulum and right-sided aortic arch: A cohort study and review of the literature. *J. Vasc. Surg.* **2004**, *39*, 131–139. [CrossRef] [PubMed]
25. Rosen, R.D.; Bordoni, B. Embryology, Aortic Arch. In *StatPearls [Internet]*; StatPearls Publishing: Treasure Island, FL, USA, 2020. Available online: https://www.ncbi.nlm.nih.gov/books/NBK553173/ (accessed on 24 June 2020).
26. Kieffer, E.; Bahnini, A.; Koskas, F. Aberrant subclavian artery: Surgical treatment in thirty-three adult patients. *J. Vasc. Surg.* **1994**, *19*, 100–111. [CrossRef]

Article

Midline and Mediolateral Episiotomy: Risk Assessment Based on Clinical Anatomy

Danielle K. Garner [1], Akash B. Patel [1], Jun Hung [1], Monica Castro [2], Tamar G. Segev [1], Jeffrey H. Plochocki [3,*] and Margaret I. Hall [2,*]

[1] Arizona College of Osteopathic Medicine, Midwestern University, Glendale, AZ 85308, USA; dgarner74@midwestern.edu (D.K.G.); apatel63@midwestern.edu (A.B.P.); jhung62@midwestern.edu (J.H.); tgsegev@gmail.com (T.G.S.)
[2] College of Graduate Studies, Midwestern University, Glendale, AZ 85308, USA; mcastr@midwestern.edu
[3] Department of Medical Education, College of Medicine, University of Central Florida, Orlando, FL 32827, USA
* Correspondence: Jeffrey.Plochocki@ucf.edu (J.H.P.); mhallx1@midwestern.edu (M.I.H.)

Abstract: Episiotomy is the surgical incision of the vaginal orifice and perineum to ease the passage of an infant's head while crowning during vaginal delivery. Although episiotomy remains one of the most frequently performed surgeries around the world, short- and long-term complications from the procedure are not uncommon. We performed midline and mediolateral episiotomies with the aim of correlating commonly diagnosed postepisiotomy complications with risk of injury to perineal neuromuscular and erectile structures. We performed 61 incisions on 47 female cadavers and dissected around the incision site. Dissections revealed that midline incisions did not bisect any major neuromuscular structures, although they did increase the risk of direct and indirect injury to the subcutaneous portion of the external anal sphincter. Mediolateral incisions posed greater risk of iatrogenic injury to ipsilateral nerve, muscle, erectile, and gland tissues. Clinician discretion is advised when weighing the potential risks to maternal perineal anatomy during vaginal delivery when episiotomy is indicated. If episiotomy is warranted, an understanding of perineal anatomy may benefit diagnosis of postsurgical complications.

Keywords: bulbs of the vestibule; midline episiotomy; mediolateral episiotomy; perineal nerve

Citation: Garner, D.K.; Patel, A.B.; Hung, J.; Castro, M.; Segev, T.G.; Plochocki, J.H.; Hall, M.I. Midline and Mediolateral Episiotomy: Risk Assessment Based on Clinical Anatomy. *Diagnostics* **2021**, *11*, 221. https://doi.org/10.3390/diagnostics11020221

Academic Editor: Heather F. Smith
Received: 20 December 2020
Accepted: 24 January 2021
Published: 2 February 2021

1. Introduction

Episiotomy is the surgical incision of the vaginal orifice and perineum to ease the passage of an infant's head while crowning during vaginal delivery. Episiotomy remains one of the most commonly performed surgeries around the world, although routine episiotomy has been on the decline since guidelines from multiple obstetric societies recommended against its use, citing insufficient evidence of its efficacy [1–4]. However, episiotomy remains an important part of the obstetrician's toolkit (even in the United States) during emergencies of fetal distress in the presence of a tight maternal perineum, especially in the case of shoulder dystocia [5,6]. However, in the same time period that routine episiotomy has fallen out of favor, obstetric and sphincter injuries, termed OASIS, have been on the rise [7–10]. OASIS is a serious maternal health concern that is associated with maternal morbidities, including pelvic floor dysfunction, fecal and urinary incontinence, sexual dysfunction, and pelvic organ prolapse [11–13]. In response to the increase in perineal injury during delivery, the prevention and management of OASIS has been deemed a priority by the international obstetric community [3,14,15].

The relationship between OASIS and episiotomy is difficult to elucidate. Episiotomy has been reported to both mitigate the risk of OASIS and be a risk factor for OASIS [2,16–18]. These conflicting findings may be explained by variations in episiotomy incisions, with midline incisions posing greater risk of perineal tearing and mediolateral incisions resulting in relatively fewer complications, especially those associated with the development of fecal continence [19–21]. It has been reported that deep perineal tears occurred in 14.8% of vaginal births using midline episiotomy, compared to only 7.0% of births with mediolateral episiotomies in the same timespan [22]. However, consistency in the placement of mediolateral incisions among practitioners has been questioned, confounding attempts to precisely evaluate the risks of episiotomy by incision location, angle, and depth [23–25]. Consequently, the relationship between diagnoses of episiotomy-related maternal morbidities and surgical incision type remain unclear, as is the role of perineal anatomic variation in birth-related injuries.

In this study, we utilize human cadaveric dissection to quantify anatomic variation in the female perineum as it relates to common episiotomy approaches. Our aim is to provide clinicians with anatomic evidence that may be used in the decision-making process to weigh the risk of injury to perineal anatomy if episiotomy is indicated during vaginal childbirth. We further aim to improve diagnoses of postepisiotomy complications by correlating perineal anatomical variation with common negative outcomes. We focus on injury risk as it relates to the origin and orientation of common episiotomy incision locations. A better understanding of perineal anatomic variation in the incisive field of episiotomy may help mitigate the risk of OASIS and other birth-related injuries. We hypothesize that each contemporary episiotomy method endangers unique perineal anatomy, and here we denote the structures at risk for these different approaches.

2. Materials and Methods

This work was funded by Midwestern University. We obtained 47 donated female human cadavers from the National Body Donor Program, St. Louis, MO, USA, a portion of the cadavers utilized by Midwestern University in medical and health sciences gross anatomy courses. All specimens were embalmed via the internal jugular vein with six gallons of embalming fluid (3% formaldehyde, 4% phenol, 31% glycerin, and 31% water). After embalming, the cadaver was stored at room temperature for a minimum of one month before delivery to Midwestern University. All cadavers were treated in accordance with local and national laws and regulations. All tissues were observed at dissection after cadaver donation was complete, and no identifying information was known for any cadaver beyond age and cause of death. Therefore, we obtained a release from the Midwestern University IRB for cadaver use in this study. We excluded any cadavers with perineal pathologies or other perineal abnormalities. The presence of scarring from past episiotomy was also noted, but due to the privacy laws associated with cadaveric donation we did not have any specific information on the obstetrical history of the donors. However, the average age of the cadavers was 72.43 with a range of 52–99, indicating that the cadavers were likely postmenopausal.

A total of 61 dissections were completed, including episiotomy incisions through superficial tissue layers were made in the midline (n = 31) and in the mediolateral direction at angles below 15° (low, n = 9), between 15 and 44° (medium, n = 10), and 45° and above (high, n = 11) prior to complete dissection of the perineum (Figure 1). We performed the incisions using standard small, sharp dissection scissors with one blade of the scissors inside the vaginal orifice and the other directly opposite outside the orifice. The majority of cadavers received both a midline and one of the mediolateral angle variants to make fuller use of the donated sample. It is important to note that the angle of the incision made during crowning does not remain constant after delivery due to anatomic changes in the vaginal wall and surrounding area [26]. The angles used in our study mirror the normal range of episiotomy scarring angles based on postpartum observations [7,16,27]. By defining incision angles in this manner, we were able to ensure that the locations of our incisions on nonpregnant cadavers coincided with those performed during crowning.

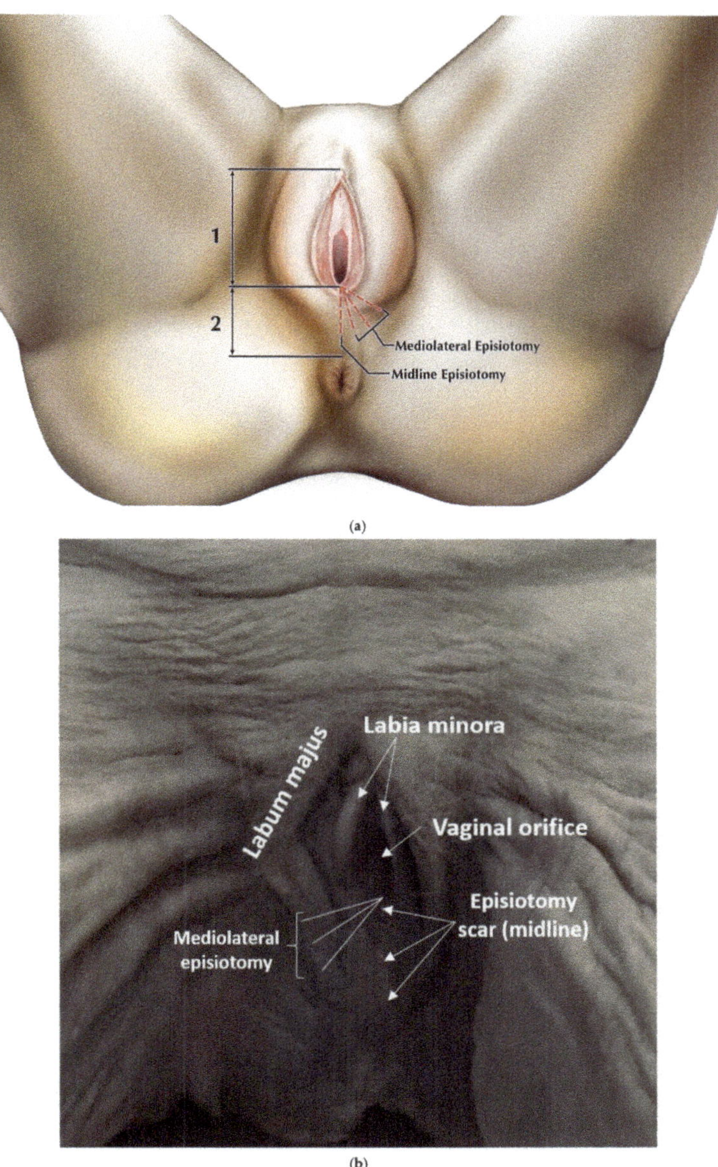

Figure 1. (**a**). Origins and angles of incisions. Incisions were made in the midline and mediolaterally at high, medium, and low angles. (**b**). Undissected female perineum, with mediolateral episiotomy incisions indicated. This specimen exhibited an already existing episiotomy scar from a midline incision. All incisions originated at the vaginal posterior fourchette.

Subsequent to incision, we measured the length of the episiotomy cuts using digital calipers to the nearest mm to ensure consistency. The incision field was then examined to identify structures transected during the procedure. We cleaned and observed nerve, muscle, erectile, and gland tissues in the incisive plane to assess injury risk. We next dissected away from the cut edges of the incision using blunt and standard dissection equipment, including scalpels, scissors, and probes where appropriate, following diagrams by Lappen [28]. We then traced the posterior labial nerves to their origin from the superficial perineal nerve, turned the cadavers prone, and followed the superficial and deep perineal nerves to their origin from the pudendal nerve. The locations of these anatomical structures were recorded for each individual in the sample and then aggregated to generate a "heat map"(Figure 5). The colors of the heat map reflect the risk of injury based on the number of structures exposed in the incisive field.

3. Results

Results of our dissections are described below, organized by episiotomy incision type, as seen in Figure 1a,b. Figure 2 displays cadaveric episiotomy incisions. Figure 3 shows the most common configuration of perineal anatomy in our sample, and Figure 4 shows one of the cadaveric dissections revealing relevant anatomy to Figure 3. Figure 5 aggregates the anatomic variation in our sample into a "heat map", with risk of injury represented by a matrix of colors. Red indicates the greatest risk, with over 80% of cadavers having major nerve, muscle, erectile, or gland structures in the incisive plane. Blue indicates low risk, with fewer than 20% of cadavers having major structures in the incisive plane.

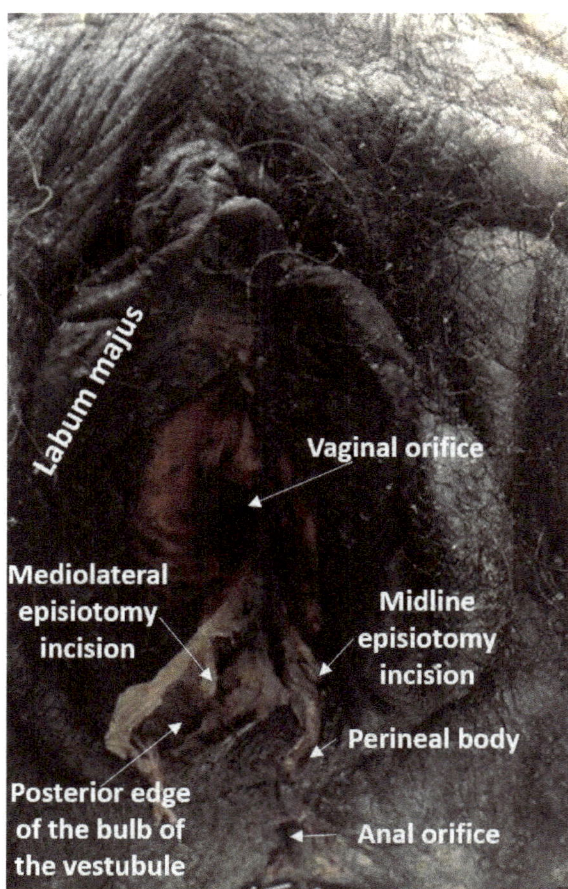

Figure 2. Female perineum with a midline episiotomy incision and a low-angle mediolateral incision. Midline episiotomies endanger the perineal body and many mediolateral incisions endanger the bulb of the vestibule and associated neurovasculature, as depicted here.

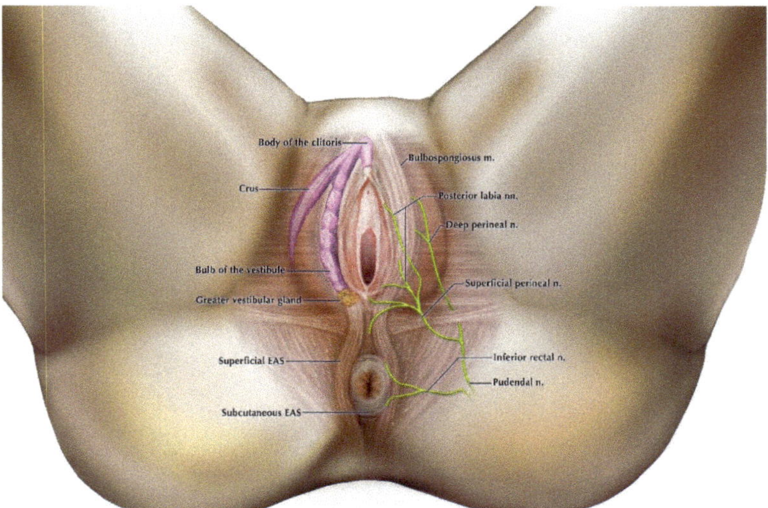

Figure 3. Representative anatomy of the perineum based on our dissections. Most commonly, the perineal body was not a major site of muscle attachment. The bulbospongiosus muscle was continuous with the superficial portion of the external anal sphincter (EAS), both innervated by branches of the perineal nerves. The inferior rectal nerve innervated the subcutaneous EAS and skin around the anus. The bulb of the vestibule extended to the posterior fourchette, with the greater vestibular gland anchored to its posterior margin.

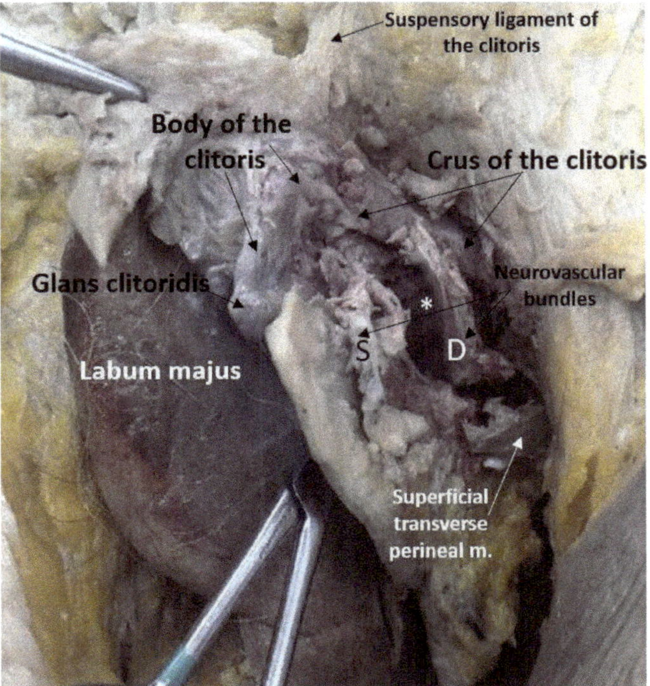

Figure 4. Female perineum dissected to reveal all perineal structures accessed by continuing a mediolateral episiotomy incision. The neurovascular bundles contain branches of the superficial (S) and deep (D) perineal nerves and the internal pudendal artery. *Bulb of the vestibule overlain by the bulbospongiosus muscle.

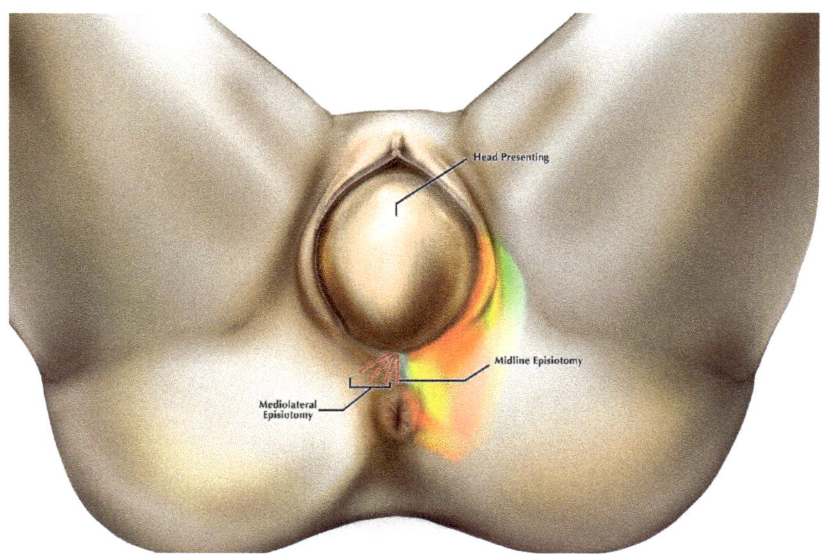

Figure 5. Heat map of risk of perineal structure injury at the time of crowning based on anatomic variation in our sample. Red-orange indicates high risk and green-blue low risk. Red dashed lines show the locations of episiotomy incisions.

3.1. Midline Incision

In 100% of cadavers, the midline incision damaged the connective tissue of the perineal body and overlying skin (see Table 1). In 16.6% of cadavers, we identified gracile muscle fibers inserting onto the perineal body, while 83.4% had only connective tissue within the incision. The predominant collagen fiber direction of the perineal body coursed parallel to the midline incision in a sagittal orientation. In 14.3% of cadavers, the incision was of sufficient length to completely bisect the perineal body and extend to some muscle fibers of the subcutaneous portion of the external anal sphincter (EAS), causing damage. However, in these instances the superficial and deep portions of the EAS were left intact, either because the majority of fibers did not converge at the midline and instead were continuous with fibers of the bulbospongiosus in the parasagittal plane, or they were sufficiently deep to avoid direct insult by the scissor blades. In the remaining 85.7% of cadavers, all portions of the EAS were undamaged by the incision. The internal anal sphincter was not directly threatened in any cadaver, nor were any major neurovascular structures located in the vicinity of the incision.

Table 1. Percentage of females with structures located in the plane of incision.

Structures in the Incisive Plane	Midline Incision $N = 31$	Mediolateral Incision $N = 30$		
	Angle: 0°	Low Angle: 10–15° $N = 9$	Medium Angle: 16–44° $N = 10$	High Angle: ≥45° $N = 11$
Perineal body	14.3%	0.0%	0.0%	0.0%
Fibers of external anal sphincter	16.6%	0.0%	0.0%	0.0%
Bulbospongiosus	0.0%	75.0%	80.0%	100%
Bulb of the vestibule	0.0%	75.0%	80.0%	100%
Deep perineal nerve branches	0.0%	25.0%	100%	40.0%
Greater vestibular gland	0.0%	75.0%	40.0%	0.0%

3.2. *High-Angle Mediolateral Incision*

In 100% of cadavers, the bulbospongiosus muscle and bulb of the vestibule were located in the plane of the incision (see Table 1). Injury to these structures was depth-dependent. Incisions greater than the combined depth of the skin and superficial fascia, which was usually no more than a few millimeters, led to muscle and erectile tissue damage. However, this thickness may not be representative of the thickness during crowning, when these structures are stretched. In 40% of cadavers, there was damage to branches of the deep perineal nerve, branches of the superficial perineal nerve, and the main trunk of the superficial perineal nerve, all of which coursed primarily in the anteromedial direction within the incision field. Therefore, the terminal branches of the superficial perineal nerve were also affected, including the posterior labial nerves, which innervate the labia minora and most of the labia majora. No direct injury to the perineal body or EAS occurred.

3.3. *Medium-Angle Mediolateral Incision*

In 100% of cadavers, posterior labial nerves coursed through the incisive plane within the superficial fascia just below the dermis of the skin and were bisected in our dissections (see Table 1). In 80% of our sample, the bulbs of the vestibule and the bulbospongiosus muscle were within the incisive plane just deeper than the superficial fascia. In total, 40% of cadavers we incised exhibited damage to the greater vestibular (Bartholin's) gland, which was consistently located in the superficial fascia and dermis near the posterior margin of the bulb of the vestibule.

3.4. *Low-Angle Mediolateral Incision*

In 75% of cadavers, the bulbospongiosus muscle, greater vestibular gland, and posterior labial nerves were in the plane of the incision. The superficial perineal nerve and its branches, as well as the bulb of the vestibule, were in the incisive plane in 25% of cadavers. The perineal body and EAS was not at risk of injury.

4. Discussion

The use of preserved cadavers for the purposes of informing surgical approaches has strengths and limitations. The formalin fixation during the embalming process produces cross-linking of proteoglycan monomers, making connective tissue stiffer than in living or fresh specimens. This has the benefit of making connective tissue structures opaque and easy to observe, as well as maintaining structures in situ during the dissection process to help maintain their anatomical location and relationships, which would not be the case with fresh samples or living anatomy during childbirth. Tissue biomechanical properties in embalmed cadavers are not representative of in vivo conditions, and therefore conclusions cannot be definitively drawn regarding the changes in anatomy that may be induced during surgery or the integrity of the tissues. However, clinical anatomical studies such as this one provide an important anatomical map to the body that would not otherwise be possible by detailing anatomical relationships that can otherwise be obscured or destroyed in living subjects. We, therefore, restrict our discussion of the findings to variations in anatomical location and structural relationships.

While the angle of the incision made during crowning does not remain constant after delivery due to anatomic changes in the perineum, careful dissections of the region can help the clinician make educated estimates of the structures that will be incised during episiotomy. In recent years, clinicians have debated the efficacy of routine episiotomy while working to define objective criteria to determine when episiotomy is indicated, including maternal perineal size, fetal size, and gestational timing, among other factors, and these discussions have led to several papers [10,14,29]. However, episiotomy is still sometimes diagnosed as medically necessary, especially when birth must be expedited in times of fetal distress during shoulder and other types of dystocia, which may be partly or wholly exacerbated by specifics of maternal perineal anatomy, usually in terms of small maternal perineal size [5,6]. In such instances, episiotomy may reduce occurrences of spontaneous perineal laceration, which is strongly correlated with pelvic organ prolapse and other complications later in life, when episiotomy alone is not [30]. In instances where spontaneous tearing does occur, it is most likely to happen in the midline [31]. In relation to this, spontaneous tearing is more common with midline episiotomy incisions in comparison to mediolateral incisions [32,33]. Our dissections implicate collagen fiber orientation in the connective tissue of the perineal body. These fibers course parallel to the midline incision in the sagittal plane, spanning the posterior fourchette and approaching the subcutaneous portion of the external anal sphincter. Their sagittal orientation places the path of least resistance in the midline, allowing spontaneous lacerations to result from stretching during labor as these fibers separate.

The small area of the perineum contains anatomy relevant to urinary, fecal, and sexual health, with no "safe" area for episiotomy where incision will not damage structure. However, our dissections confirmed that the perineal body was not a prominent site of muscle attachment in the majority of females we studied. This finding has been reported in other anatomical and histological investigations that describe the perineal body as having little or no insertion of striated fibers of the external anal sphincter or bulbospongiosus into the perineal body [33–35], and therefore it may not provide substantial protection from tearing during delivery. The expectation that the perineal body provides a major site of muscle attachments in all females selected for episiotomy may bias decisions about where to perform episiotomy, as well as surgical techniques on perineum reconstruction should tearing occur. Rather than considering the bulbospongiosus muscles and the superficial external anal sphincter as discrete, circular muscles with a common attachment to the anterior and posterior portions of the perineal body, our dissections agree with previous observations that anatomically these muscles typically comprise a single, continuous sling surrounding both the vaginal orifice and the external anal orifice that does not attach at the midline. Both portions of this sling share a common innervation from the superficial perineal nerve, which approaches the musculature laterally (Figures 3 and 4) [36]. Therefore, midline episiotomy incisions would not pose a serious risk to these neuromuscular structures in the females we studied. Conversely, however, since the perineal body lacked strong muscular support in our sample of females, midline episiotomy may lead to statistically larger numbers of third- and fourth-degree tears. In addition, we found considerable variation in the distance from the posterior fourchette to the anal orifice, indicating the size of the perineal body. Incision lengths in our study were comparable to those performed in surgery [15] and consistently these at least partially bisected the perineal body. These factors are significant because the perineal body length is a large risk factor that is negatively correlated with spontaneous laceration [37,38]. Midline episiotomy in females with shorter perinea warrants extreme caution or should perhaps be avoided altogether.

In comparison, our dissections find that mediolateral episiotomy approaches do not endanger the perineal body or the subcutaneous external anal sphincter directly. Mediolateral incisions were oblique to connective tissue fiber orientation in our dissections. Our observations corroborate the results from studies of patient outcomes showing that mediolateral incisions during crowning protect against postpartum fecal incontinence from third- and fourth-degree perineal tearing by diverting forces away from the subcutaneous external anal sphincter and the perineal body [39–41]. However, indirect injury to the superficial external anal sphincter that is not visually evident may be diagnosed sonographically and is significant enough to elicit complaints of postpartum fecal incontinence [42,43]. We find that mediolateral incisions jeopardize the bulbospongiosus muscle, which as discussed above extends posteriorly beyond the bulb of the vestibule to be continuous with fibers of the superficial external anal sphincter to create a sling encompassing both the vaginal and external anal orifices. Anatomically, injury to these fibers may affect fecal continence by causing asymmetric contractions around the anal canal. Endoanal ultrasound studies often display this asymmetric tearing pattern [43,44]. Additionally, bulbospongiosus and the superficial external anal sphincter share innervation [34,45]. The functionality of the external anal sphincter may be impaired if nerve damage, either through incision or traction, occurs.

The risk of nerve damage is greater in mediolateral incision because the perineal nerve and its branches course lateral to midline, while midline episiotomy does not anatomically endanger the superficial or deep EAS. Even if the subcutaneous EAS is injured in midline incision, this gracile muscle will heal with its nerve supply intact, and therefore will not affect long-term fecal continence. These anatomic findings should be considered when contemplating episiotomy approaches.

We also find that mediolateral incisions place several important structures related to sexual health at greater risk for injury in comparison to the midline incision, including the bulbs of the vestibule, bulbospongiosus muscle, vestibular gland, branches of the deep perineal nerve, and the trunk of the superficial perineal nerve, along with its terminal branches, the posterior labial nerves. The incisive plane endangered these structures, which serve important sexual functions, in one or more of the mediolateral incision angles. The skin in this region is well innervated and contains more mechanosensory Merkel's cells than any other epithelium in the body [46]. The bulbs of the vestibule are the physiological seat of female orgasm, and they, along with the labia minora, swell during arousal and orgasm [46–48]. Greater vestibular glands are the source of preorgasmic vaginal secretions that contribute to lubrication and protection of the vagina during intercourse [49]. The bulbospongiosus–superficial external anal sphincter muscular complex contracts during sexual arousal to maintain blood within the bulbs of the vestibule, which contributes to clitoral erection, as well as providing the contractions of orgasm [34,35]. Women with midline episiotomies report both a shorter time before they engage in sexual activity than those that receive mediolateral episiotomies, and they also report no change in postpartum orgasm number; it may be that direct or indirect injury to these sexual structures is implicated in changes in postpartum sexual behavior [50,51]. Discussions of sexual activity have historically been avoided in postpartum examinations, leading to the suspected underreporting of sexual dysfunction [52,53]. Additional study is needed to evaluate the effects of episiotomy on sexual anatomy and function.

Recovery time from injury to perineal anatomy is another factor that should be considered when episiotomy is indicated. However, most surveys investigating parturition outcomes collect data relatively soon after birth, and long-term functional deficits may fail to be noted. One study that followed up with mothers at three and six months postpartum found that fecal incontinence was elevated three months after perineal tearing following midline episiotomy, but by six months the differences in fecal continence were no longer statistically significant [54]. A 10-year follow-up study showed that rates of fecal incontinence were similar between those who had perineal tearing during delivery and those who did not, regardless of whether episiotomy was performed [55]. These data seem to suggest the episiotomy type may not differ in their long-term consequences, but it is unclear if additional data from larger studies would confirm this same pattern. It is also unclear how other complications, such as urinary incontinence, sexual dysfunction, and pelvic organ prolapse, differ in postpartum years with episiotomy type, as sufficient data are lacking. Nonetheless, clinicians should incorporate the risk of short- and long-term injury into the decision to perform midline versus mediolateral episiotomy accordingly. Similarly, knowledge of perineal anatomy should be applied when diagnosing postepisiotomy complications.

5. Conclusions

We find that midline and mediolateral episiotomy incisions each pose unique risks to perineal anatomy, and there is no incision site that does not endanger structure. A better understanding of these risks and of the relevant anatomy is important, as clinician knowledge of perineal anatomy has been reported to be "suboptimal" and may affect individual approaches to reducing the risk of obstetric perineal injury [56]. In fact, most clinicians perform the type of episiotomy they learned in postgraduate training [21], which in the United States is the midline approach and in Europe is the mediolateral approach [26,27,57], rather than tailoring their incision to the unique circumstance of their patient. Risk to perineal anatomy should be part of the decision-making process, as should short- and long-term risks to fecal continence and sexual health. Clinician discretion is needed when balancing the risks to maternal perineal anatomy during vaginal delivery when considering episiotomy. A complete knowledge of perineal anatomy also aids the diagnosis of OASIS and other complications related to episiotomy incisions.

Author Contributions: M.I.H. and J.H.P. conceived of the original manuscript. A.B.P. and T.G.S. researched and designed the dissection approaches. M.I.H., J.H.P., T.G.S., A.B.P., and J.H. together designed the dissections. M.I.H., J.H.P., A.B.P., J.H., D.K.G., and M.C. carried out and interpreted dissections. D.K.G. and A.B.P. wrote the initial manuscript. M.I.H. and J.H.P. revised the final manuscript. All authors have read and agreed to the published version of the manuscript.

Funding: This project was funded by Midwestern University Intramural Funds and AZCOM Research Fellowships to D.K.G., A.B.P., and J.H. Funding did not play any role in the design of the study; in the collection, analysis, or interpretation of data; or in the writing of the manuscript.

Acknowledgments: As succinctly put in Iwanga et al. 2020, "The authors sincerely thank those who donated their bodies to science so that anatomical research could be performed. Results from such research can potentially increase mankind's overall knowledge that can then improve patient care. Therefore, these donors and their families deserve our highest gratitude" [58]. We also thank Heather Smith for useful discussions, access to specimens, age data, and the invitation to submit; Ashley Bergeron for facilitating dissection; and Ryan Dickerson for providing the illustrations.

Conflicts of Interest: The authors declare that they have no conflicts of interest.

References

1. American College of Obstetricians and Gynecologists. ACOG Practice Bulletin No. 71: Episiotomy. *Obstet. Gynecol.* **2006**, *107*, 956–962. [CrossRef]
2. American College of Obstetricians and Gynecologists. Practice Bulletin No. 165: Prevention and Management of Obstetric Lacerations at Vaginal Delivery. *Obstet. Gynecol.* **2016**, *128*, e1–e15. [CrossRef] [PubMed]
3. Cargill, Y.M.; MacKinnon, C.J. SOGC Clinical Practice Guidelines, No. 148, August 2004. *J. Obstet. Gynaecol. Can.* **2004**, *26*, 747–753. [PubMed]
4. Royal College of Obstetricians and Gynaecologists. *Operative Vaginal Delivery*; Green-Top Guideline No. 26; Royal College of Obstetricians and Gynaecologists: London, UK, 2011.
5. Sagi-Dain, L.; Sagi, S. The role of episiotomy in prevention and management of shoulder dystocia: A systematic review. *Obstet. Gynecol. Surv.* **2015**, *70*, 354–362. [CrossRef] [PubMed]
6. Hartmann, K.; Viswanathan, M.; Palmieri, R.; Gartlehner, G.; Thorp, J., Jr.; Lohr, K.N. Outcomes of routine episiotomy: A systematic review. *JAMA* **2005**, *293*, 214148. [CrossRef]
7. Dahlen, H.G.; Priddis, H.; Thornton, C. Severe perineal trauma is rising, but let us not overreact. *Midwifery* **2015**, *31*, 1–8. [CrossRef]
8. Ekeus, C.; Nilsson, E.; Gottvall, K. Increasing incidence of anal sphincter tears among primiparas in Sweden: A population-based register study. *Acta Obstet. Gynecol. Scand.* **2008**, *87*, 564–573. [CrossRef]
9. Gurol-Urganci, I.; Cromwell, D.A.; Edozien, L.C.; Mahmood, T.A.; Adams, E.J.; Richmond, D.H.; Templeton, A.; van der Meulen, J.H. Third- and fourth-degree perineal tears among primiparous women in England between 2000 and 2012: Time trends and risk factors. *BJOG* **2013**, *120*, 1516–1525. [CrossRef]
10. Ismail, S.I.; Puyk, B. The rise of obstetric anal sphincter injuries (OASIS): 11-year trend analysis using Patient Episode Database for Wales (PEDW) data. *Obstet. Gynecol.* **2014**, *34*, 495–498. [CrossRef]
11. Mous, M.; Muller, S.; de Leeuw, J. Long-term effects of anal sphincter rupture during vaginal delivery: Faecal incontinence and sexual complaints. *BJOG* **2008**, *115*, 234–238. [CrossRef]
12. Samarasekera, D.N.; Bekhit, M.T.; Wright, Y.; Lowndes, R.H.; Stanley, K.P.; Preston, J.P.; Preston, P.; Speakman, C.T.M. Long-term anal continence and quality of life following postpartum anal sphincter injury. *Colorectal Dis.* **2008**, *10*, 793–799. [CrossRef] [PubMed]
13. van Brummen, H.; Bruinse, H.; van de Pol, G.; Heintz, A.; van der Vaart, G. Which factors determine the sexual function 1 year after childbirth? *BJOG* **2006**, *113*, 914–918. [CrossRef] [PubMed]
14. Harvey, M.A.; Pierce, M. Obstetrical anal sphincter injuries (OASIS): Prevention, recognition, and repair. SOGC Clinical Practice Guideline, No. 330. *J. Obstet. Gynaecol. Can.* **2015**, *37*, 1131–1149. [CrossRef]
15. Royal College of Obstetricians and Gynaecologists. *Third- and Fourth-Degree Perineal Tears, Management*; Green Top Guideline No. 29; Royal College of Obstetricians and Gynaecologists: London, UK, 2017.
16. Gonzalez-Díaz, E.; Cea, L.M.; Corona, A.F. Trigonometric characteristics of episiotomy and risks for obstetric anal sphincter injuries in operative vaginal delivery. *Int. Urogynecol. J.* **2015**, *26*, 235–242. [CrossRef] [PubMed]

17. Lund, N.S.; Persson, L.K.; Jangö, H.; Gommesen, D.; Westergaard, H.B. Episiotomy in vacuum-assisted delivery affects the risk of obstetric anal sphincter injury: A systematic review and meta-analysis. *Eur. J. Obstet. Gynecol. Reprod. Biol.* **2016**, *207*, 193–199. [CrossRef] [PubMed]
18. van Bavel, J.; Hukkelhoven, C.W.; de Vries, C.; Papatsonis, D.N.; de Vogel, J.; Roovers, J.P.; Mol, B.W.; de Leeuw, J.W. The effectiveness of mediolateral episiotomy in preventing obstetric anal sphincter injuries during operative vaginal delivery: A ten-year analysis of a national registry. *Int. Urogynecol. J.* **2018**, *29*, 407–413. [CrossRef]
19. Correa, M.D.; Passini, R. Selective episiotomy: Indications, technique, and association with severe perineal lacerations. *Rev. Bras. Ginecol. Obstet.* **2016**, *38*, 301–307. [CrossRef]
20. Eogan, M.; Daly, L.; O'connell, P.R.; O'herlihy, C. Does the angle of episiotomy affect the incidence of anal sphincter injury? *BJOG* **2006**, *113*, 190–194.
21. Menzies, R.; Leung, M.; Chandrasekaran, N.; Lausman, A.; Geary, M. Episiotomy Technique and Management of Anal Sphincter Tears—A Survey of Clinical Practice and Education. *J. Obstet. Gynaecol. Can.* **2016**, *38*, 1091–1099. [CrossRef]
22. Sooklim, R.; Thinkhamrop, J.; Lumbiganon, P.; Prasertcharoensuk, W.; Pattamadilok, J.; Seekorn, K.; Chongsomchai, C.; Pitak, P.; Chansamak, S. The outcomes of midline versus medio-lateral episiotomy. *Reprod. Health* **2007**, *4*, 10. [CrossRef]
23. Andrews, V.; Thakar, R.; Sultan, A.H.; Jones, P.W. Are mediolateral episiotomies actually mediolateral? *BJOG* **2005**, *112*, 1156–1158. [CrossRef] [PubMed]
24. Silf, K.; Woodhead, N.; Kelly, J.; Fryer, A.; Kettle, C.; Ismail, K.M. Evaluation of accuracy of mediolateral episiotomy incisions using a training model. *Midwifery* **2015**, *31*, 197–200. [CrossRef] [PubMed]
25. Wong, K.W.; Ravindran, K.; Thomas, J.M.; Andrews, V. Mediolateral episiotomy: Are trained midwives and doctors approaching it from a different angle? *Eur. J. Obstet. Gynecol. Reprod. Biol.* **2014**, *174*, 46–50. [CrossRef] [PubMed]
26. Kalis, V.; Karbanova, J.; Horak, M.; Lobovsky, L.; Kralickova, M.; Rokyta, Z. The incision angle of mediolateral episiotomy before delivery and after repair. *Int. J. Gynecol. Obstet.* **2008**, *103*, 5–8. [CrossRef] [PubMed]
27. Kalis, V.; Stepan, J., Jr.; Horak, M.; Roztocil, A.; Kralickova, M.; Rokyta, Z. Definitions of mediolateral episiotomy in Europe. *Int. J. Gynecol. Obstet.* **2008**, *100*, 188–189.
28. Lappen, J.R.; Isaacs, M. *Episiotomy and Repair Technique*; 2016. Available online: https://emedicine.medscape.com/article/2047173-technique (accessed on 9 July 2019).
29. Friedman, A.M.; Ananth, C.V.; Prendergast, E.; D'alton, M.E.; Wright, J.D. Variation in and factors associated with use of episiotomy. *JAMA* **2015**, *313*, 197–199. [CrossRef]
30. Handa, V.L.; Blomquist, J.L.; McDermott, K.C.; Friedman, S.; Muñoz, A. Pelvic floor disorders after vaginal birth: Effect of episiotomy, perineal laceration, and operative birth. *Obstet. Gynecol.* **2012**, *119*, 233–239. [CrossRef]
31. Thacker, S.B. Midline versus mediolateral episiotomy: We still don't know which cut is better or how beneficial the procedure is. *BMJ* **2000**, *320*, 1615–1616. [CrossRef]
32. Kudish, B.; Blackwell, S.; Mcneely, S.G.; Bujold, E.B.; Kruger, M.; Hendrix, S.L.; Sokol, R. Operative vaginal delivery and midline episiotomy: A bad combination for the perineum. *AJOG* **2006**, *195*, 749–754. [CrossRef]
33. Shionio, P.; Klebaof, M.A.; Carey, J.C. Midline episiotomies: More harm than good? *Obstet. Gynecol.* **1990**, *75*, 765–770.
34. Plochocki, J.H.; Rodriguez-Sosa, J.R.; Adrian, B.; Ruiz, S.A.; Hall, M.I. A functional and clinical reinterpretation of human perineal neuromuscular anatomy: Application to sexual function and continence. *Clin. Anat.* **2016**, *29*, 1053–1058. [CrossRef]
35. Shafik, A.; El Sibai, O.; Shafik, A.A.; Shafik, I.A. A novel concept for the surgical anatomy of the perineal body. *Dis. Colon Rectum* **2007**, *50*, 2120–2125. [CrossRef] [PubMed]
36. Soga, H.; Nagata, J.; Murakami, G.; Yajima, T.; Takenaka, A.; Fujisawa, M.; Koyama, M. A histotopographic study of the perineal body in elderly women: The surgical applicability of novel histological findings. *Int. Urogynecol. J.* **2007**, *18*, 1423–1430. [CrossRef] [PubMed]
37. Geller, E.J.; Robinson, B.L.; Matthews, C.A.; Celauro, K.P.; Dunivan, G.C.; Crane, A.K.; Ivins, A.R.; Woodham, P.C.; Fielding, J.R. Perineal body length as a risk for ultrasound-diagnosed anal sphincter tear at first delivery. *Int. Urogynecol. J.* **2014**, *25*, 631–636. [CrossRef] [PubMed]
38. Kalis, V.; Chaloupka, P.; Turek, J.; Rokyta, Z. The perineal body length and injury at delivery. *Ceska Gynekol.* **2005**, *70*, 355–361.
39. Fernando, R.J.; Sultan, A.H.; Freeman, R.H.; Williams, A.A.; Adams, E.A. The management of third- and fourth-degree perineal tears. *Green Top. Guideline No. 29.* London: Royal College of Obstetricians and Gynaecologists. *BJOG* **2015**, *6*, 1–19.
40. Nygaard, I.E.; Rao, S.S.; Dawson, J.D. Anal incontinence after anal sphincter disruption: A 30-year retrospective cohort study. *Obstet. Gynecol.* **1997**, *89*, 896–901.
41. Tincello, D.G.; Williams, A.; Fowler, G.E.; Adams, E.J.; Richmond, D.H.; Alfirevic, Z. Differences in episiotomy technique between midwives and doctors. *BJOG* **2003**, *110*, 1041–1044. [CrossRef]
42. Andrews, V.; Thakar, R.; Sultan, A.H.; Jones, P.W. Occult anal sphincter injuries—Myth or reality? *BJOG* **2006**, *113*, 195–200. [CrossRef]
43. Staric, K.D.; Lukanovic, A.; Petrocnik, P.; Zacesta, V.; Cescon, C.; Lucovnik, M. Impact of mediolateral episiotomy on incidence of obstetrical anal sphincter injury diagnosed by endoanal ultrasound. *Midwifery* **2017**, *51*, 40–43. [CrossRef]
44. Sudoł-Szopinńska, I.; Radkiewicz, J.; Szopiński, T.; Panorska, A.K.; Jakubowski, W.; Kawka, J. Postpartum endoanal ultrasound findings in primiparous women after vaginal delivery. *Acta Radiol.* **2010**, *51*, 819–824. [CrossRef]

45. Hall, M.I.; Rodriguez-Sosa, J.R.; Plochocki, J.R. Reorganization of mammalian body wall patterning with cloacal septation. *Sci. Rep.* **2017**, *7*, 9182. [CrossRef] [PubMed]
46. Neill, S.M.; Lewis, F.M. *Ridley's the Vulva;* Wiley-Blackwell: Hoboken, NJ, USA, 2009.
47. Pauls, R.N. Anatomy of the clitoris and the female sexual response. *Clin. Anat.* **2015**, *28*, 376–384. [CrossRef] [PubMed]
48. Puppo, V. Anatomy and physiology of the clitoris, vestibular bulbs, and labia minora with a review of the female orgasm and the prevention of female sexual dysfunction. *Clin. Anat.* **2013**, *26*, 134–152. [CrossRef] [PubMed]
49. Masters, W.H.; Johnson, V.E. *Human Sexual Response;* Little Brown: Boston, MA, USA, 1966.
50. Coats, P.M.; Chan, K.K.; Wilkins, M.; Beard, R.J. A comparison between midline and mediolateral episiotomies. *BJOG* **1980**, *87*, 408–412. [CrossRef] [PubMed]
51. Connolly, A.; Thorp, J.; Pahel, L. Effects of pregnancy and childbirth on postpartum sexual function: A longitudinal prospective study. *Int. Urogynecol. J.* **2005**, *16*, 263–267. [CrossRef]
52. Glazener, C.M.A. Sexual function after childbirth: Women's experiences, persistent morbidity and lack of professional recognition. *Br. J. Obstet. Gynaecol.* **1997**, *104*, 330–333. [CrossRef]
53. Ergenoglu, A.M.; Yeniel, A.Ö.; Itil, I.M.; Askar, N.; Meseri, R.; Petri, E. Overactive bladder and its effects on sexual dysfunction among women. *Acta Obstet. Gynecol. Scand.* **2013**, *92*, 1202–1207. [CrossRef]
54. Signorello, L.B.; Harlow, B.L.; Chekos, A.K.; Repke, J.T. Midline episiotomy and anal incontinence: Retrospective cohort study. *BMJ* **2000**, *320*, 86–90. [CrossRef]
55. Faltin, D.L.; Otero, M.; Petignat, P.; Sangalli, M.R.; Floris, L.A.; Boulvain, M.; Irion, O. Women's health 18 years after rupture of the anal sphincter during childbirth: I. Fecal incontinence. *AJOG* **2006**, *194*, 1255–1259. [CrossRef]
56. Sultan, A.H.; Kamm, M.A.; Hudson, C.N. Obstetric perineal tears: An audit of training. *J. Obstet. Gynaecol.* **1995**, *15*, 19–23. [CrossRef]
57. Frankman, E.A.; Wang, L.; Bunker, C.H.; Lowder, J.L. Episiotomy in the United States: Has anything changed? *AJOG* **2009**, *200*, 573.e1–573.e7. [CrossRef] [PubMed]
58. Iwanga, I.; Singh, V.; Ohtsuka, A.; Hwang, Y.; Kim, H.-J.; Morys, J.; Ravi, K.S.; Ribatti, D.; Trainor, P.A.; Sanudo, J.R.; et al. Acknowledging the use of human cadaveric tissues in research papers: Recommendations from anatomical journal editors. *Clin. Anat.* **2021**, *34*, 2–4. [CrossRef] [PubMed]

MDPI

Review

Pelvic Lymphadenectomy in Gynecologic Oncology—Significance of Anatomical Variations

Stoyan Kostov [1], Yavor Kornovski [1], Stanislav Slavchev [1], Yonka Ivanova [1], Deyan Dzhenkov [2], Nikolay Dimitrov [3] and Angel Yordanov [4,*]

1 Department of Gynecology, Medical University Varna "Prof. Dr. Paraskev Stoyanov", 9002 Varna, Bulgaria; drstoqn.kostov@gmail.com (S.K.); ykornovski@abv.bg (Y.K.); st_slavchev@abv.bg (S.S.); yonka.ivanova@abv.bg (Y.I.)
2 Department of General and Clinical Pathology, Forensic Medicine and Deontology, Division of General and Clinical Pathology, Faculty of Medicine, Medical University Varna "Prof. Dr. Paraskev Stoyanov", 9002 Varna, Bulgaria; dzhenkov@mail.bg
3 Department of Anatomy, Faculty of Medicine, Trakia University, 6000 Stara Zagora, Bulgaria; nikolay.dimitrov@trakia-uni.bg
4 Department of Gynecologic Oncology, Medical University Pleven, 5800 Pleven, Bulgaria
* Correspondence: angel.jordanov@gmail.com

Abstract: Pelvic lymphadenectomy is a common surgical procedure in gynecologic oncology. Pelvic lymph node dissection is performed for all types of gynecological malignancies to evaluate the extent of a disease and facilitate further treatment planning. Most studies examine the lymphatic spread, the prognostic, and therapeutic significance of the lymph nodes. However, there are very few studies describing the possible surgical approaches and the anatomical variations. Moreover, a correlation between anatomical variations and lymphadenectomy in the pelvic region has never been discussed in medical literature. The present article aims to expand the limited knowledge of the anatomical variations in the pelvis. Anatomical variations of the ureters, pelvic vessels, and nerves and their significance to pelvic lymphadenectomy are summarized, explained, and illustrated. Surgeons should be familiar with pelvic anatomy and its variations to safely perform a pelvic lymphadenectomy. Learning the proper lymphadenectomy technique relating to anatomical landmarks and variations may decrease morbidity and mortality. Furthermore, accurate description and analysis of the majority of pelvic anatomical variations may impact not only gynecological surgery, but also spinal surgery, urology, and orthopedics.

Keywords: anatomical landmarks; anatomical variations; pelvic lymph nodes; gynecologic oncology; pelvic lymphadenectomy

Citation: Kostov, S.; Kornovski, Y.; Slavchev, S.; Ivanova, Y.; Dzhenkov, D.; Dimitrov, N.; Yordanov, A. Pelvic Lymphadenectomy in Gynecologic Oncology—Significance of Anatomical Variations. *Diagnostics* **2021**, *11*, 89. https://doi.org/10.3390/diagnostics11010089

Received: 28 November 2020
Accepted: 5 January 2021
Published: 7 January 2021

1. Introduction

Pelvic lymph node dissection (PLND) is a common surgical procedure in gynecologic oncology [1]. The lymphatic system is the primary dissemination pathway for gynecological malignancies. PLND is applied for cancer staging, prognosis, surgical, and postoperative management [2,3]. PLND is performed for all types of gynecological malignancies to evaluate the extent of a disease and facilitate further treatment planning. Additionally, PLND is beneficial in cases where removing metastatic lymph nodes improves overall survival and disease-free survival [4]. Most studies examine the lymphatic spread, the prognostic, and therapeutic significance of pelvic lymph nodes. However, there are very few studies describing the possible surgical approach, dissection techniques and anatomical variations [5]. There is limited information and disagreement on lymph nodes location, groups, and overall number [6].

Moreover, a correlation between anatomical variations and PLND in the pelvic region has never been discussed in medical literature. Surgeons should be familiar with pelvic anatomy and its variations to safely perform PLND. Learning the proper lymphadenectomy technique relating to anatomical landmarks and variations may decrease morbidity and mortality [7]. The present article aims to define, detail, and summarize the anatomic landmarks during PLND in gynecologic oncology. Furthermore, a summary of the most common anatomical variations (of nerves, vessels, ureters) and potential complications related to PLND in the pelvic region are clearly defined.

2. Pelvic Lymph Nodes and Regions

Knowledge of the anatomical localization of lymph node groups in the pelvis is essential for the effectiveness and safety of lymphadenectomy [8]. The pelvic lymph nodes and the connecting lymphatic channels communicate with the venous system; the lymphatic system embryologically develops from vascular plexuses, arising from the venous system [3]. The anatomical localization of the major groups and sub-groups of pelvic lymph nodes is summarized in Figures 1–3 [2,6,8–15].

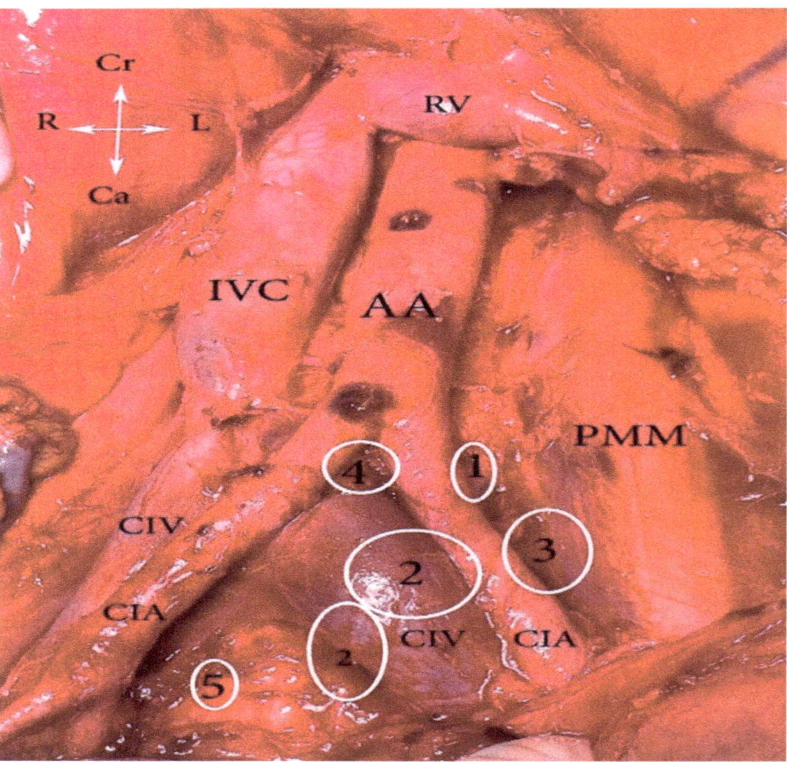

Figure 1. Common iliac lymph nodes classification (open surgery). **1.** Lateral—between lateral part of CIV and medial part of psoas major muscle, **2.** medial—medial to CIV and CIA, **3.** middle—located in the lumbosacral fossa, **4.** subaortic—below aortic bifurcation, **5.** promontory—at the promontory. AA—abdominal aorta, IVC—inferior vena cava, RV—right renal vein, PMM—psoas major muscle, CIA—common iliac artery, CIV—common iliac vein, Cr—cranial, Ca—caudal, L—Left, R—right.

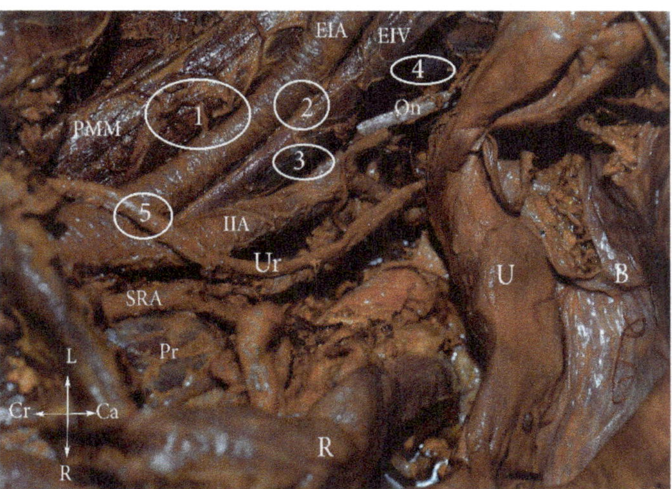

Figure 2. External iliac lymph nodes classification (embalmed cadaver). **1.** Lateral—lateral to external iliac artery, **2.** middle—medial to the EIA and lateral to the EIV, **3.** medial—medial to both external iliac vessels, **4.** obturator—around the obturator nerve and vessels, **5.** interiliac—at the level of CIA bifurcation, between the EIA and IIA. PMM—psoas major muscle, EIA—external iliac artery, EIV—external iliac vein, IIA—internal iliac artery, Ur—ureter, U—uterus, B—bladder, SRA—superior rectal artery, Pr—promontorium, R—rectum, L—left, r—right, Cr—cranial, Ca—caudal.

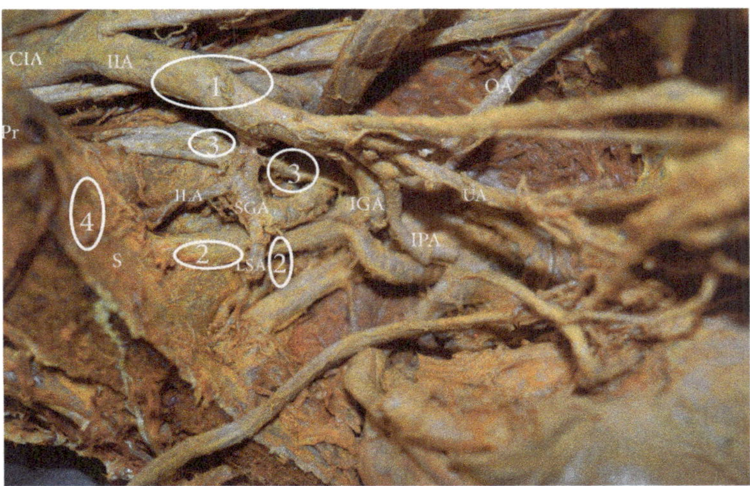

Figure 3. Internal iliac lymph nodes classification (embalmed cadaver—left hemipelvis). **1.** Anterior—anterior to anterior division of internal iliac artery, **2.** lateral sacral—close to the paired lateral sacral arteries, **3.** gluteal—between superior gluteal and internal iliac artery, **4.** sacral (presacral)—along median sacral artery. CIA—common iliac artery, IIA—internal iliac artery, OA—obturator artery, UA—umbilical artery, IPA—internal pudendal artery, IGA—inferior gluteal artery, LSA—lateral sacral artery, ILA—iliolumbar artery, Pr—promontorium, S—sacrum.

Some authors include the subaortic and promontory lymph nodes in the medial common iliac lymph group; the obturator lymph nodes may also be considered as part of the medial external lymph group. Moreover, in some articles, anterior, lateral sacral, and gluteal internal lymph nodes are defined as junctional lymph nodes [8,10,12].

We present an anatomical classification of the pelvic lymph nodes rather than a clinical one. Although the anatomical classification is rather complex, it better defines the localization of the pelvic lymph nodes. Furthermore, the anatomical variations, significant for the PLND, are better described.

In medical literature and among surgeons, there are large variations and discrepancies in the nomenclature of the pelvic lymph node regions [2]. The present article and the majority of authors recognize the following pelvic lymph node regions: common iliac; external iliac; internal iliac; obturator; and sacral (or presacral) [5,7,12,13]. Some authors add parametrial, mesorectal, and interiliac regions, while others include tissue from the interiliac region to the external iliac and obturator regions and remove the tissue from the parametrial region together with the parametrium during radical hysterectomy [5,14,15].

3. Selective PLND—Anatomical Landmarks and Techniques

There are several surgical procedures related to dissection of pelvic lymph nodes: sentinel lymph node biopsy, pelvic lymph node sampling, selective lymphadenectomy and complete (systematic) PLND. The present article describes the systematic lymphadenectomy in which all pelvic lymph nodes, draining the pelvic organs, have been removed [1]. Cibula and Rustum presented detailed anatomic boundaries for five pelvic lymph node regions: external iliac, obturator, internal iliac, common iliac, and presacral region [5]. Anatomic boundaries of the common PLND are: ventral—common iliac artery bifurcations, dorsal—abdominal aorta bifurcation, lateral—psoas major muscle, medial—right-medial aspect of common iliac vessels, left-mesoureter [5]. Anatomic boundaries of the external and the internal PLND include common iliac artery bifurcation dorsally, the deep circumflex iliac vein ventrally, and the genitofemoral nerve of the psoas major muscle laterally. The medial border is the obliterated umbilical artery ventrally and the ureter dorsally (Figures 4 and 5) [1,7,15].

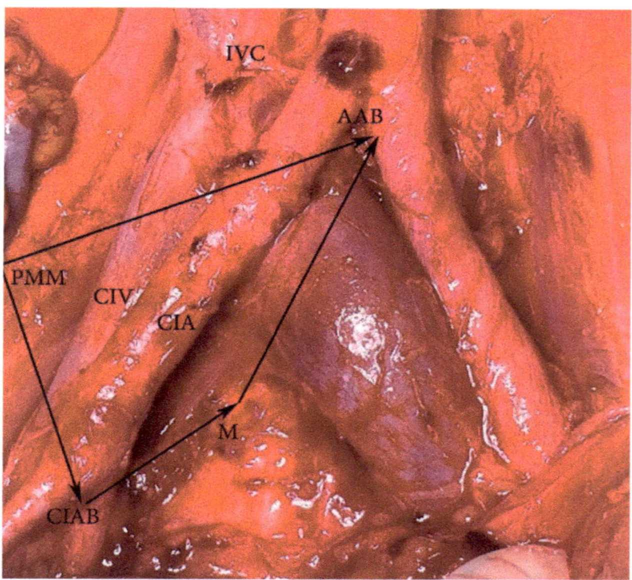

Figure 4. Common iliac lymph nodes dissection—anatomic boundaries (open surgery, right side). Dorsal—abdominal aorta bifurcation, ventral—common iliac artery bifurcation, medial—medial aspect of common iliac vessels (on the left side is mesoureter), lateral—psoas major muscle. AAB—abdominal aorta bifurcation, PMM—psoas major muscle, CIAB—common iliac artery bifurcation, M—medial aspect of common iliac vessels on the right side (on the left side is mesoureter), IVC—inferior vena cava, CIA—common iliac artery, CIV—common iliac vein.

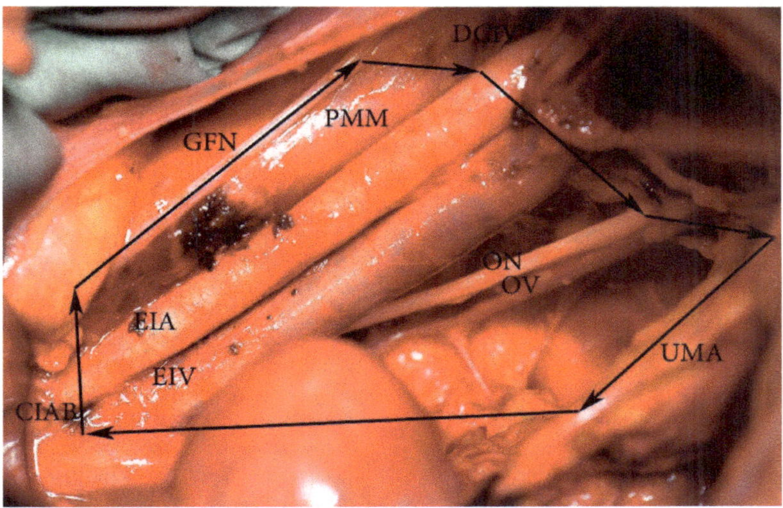

Figure 5. External and internal iliac lymph nodes dissection—anatomic boundaries (open surgery left pelvic sidewall). CIAB—common iliac artery bifurcation, ON—obturator nerve, DCIV—deep circumflex iliac vein, GFN—genitofemoral nerve, PMM—psoas major muscle, EIA—external iliac artery, EIV—external iliac vein, OV—obturator vein. The medial border is the ureter dorsally and obliterated umbilical artery ventrally. In the figure, the ureter is stretched medially for better exposure of the visible structures.

4. Systematic Open PLND—Surgical Technique and Steps

(1) Peritoneal incision. After transecting (not a necessary step) the round ligament, the peritoneum is incised in a posterior (lateral and parallel to the infundibulopelvic ligament) and an anterior (ventrally and laterally to the obliterated umbilical artery) direction. The iliac vessels are exposed from the bifurcation of the aorta to the inguinal ligament.

(2) Identification of the ureter.

(3) Lateral paravesical and lateral (Latzko's space) pararectal space dissection. The lateral paravesical space is dissected between the obliterated umbilical artery and the external iliac vessels. Lateral pararectal space is dissected between the internal iliac artery and the ureter.

(4) Genitofemoral nerve identification. Lateral incision to the fascia of the psoas muscle is preferable in order to avoid genitofemoral nerve injury. The nerve is located lateral to the external iliac vessels and sometimes overlying them.

(5) External iliac region dissection: lateral and medial external iliac vessels dissection. The dissection begins at the origin of the external iliac vessels and finishes down to the point where the deep circumflex iliac vein crosses over the external iliac artery.

(6) Obturator region dissection. The obturator space is approached by retracting the external iliac vessels medially and the psoas muscle laterally, and by dissection of the areolar tissue that lies directly between these vessels and the lateral pelvic wall. The obturator nerve is identified. The procedure is followed by lateral retraction of the external iliac vessels to expose the obturator space. Superficial obturator lymph nodes are dissected after obturator nerve visualization (obturator nerve stripping). For locally advanced cervical cancer cases, PLND continues with dissection of the deep obturator lymph nodes and the gluteal nodes.

(7) Internal iliac region dissection. Lymph nodes are removed medially and anteriorly to the internal iliac vessels.

(8) Common iliac region dissection. Lymph nodes are removed ventrally and laterally from both common iliac vessels to the aortic bifurcation. Middle common iliac lymph nodes are located in the lumbosacral fossa. It is approached by medial retraction of the common iliac vessels and lateral retraction of the psoas muscle. The obturator nerve, entering the obturator fossa through the body of the psoas muscle, the iliolumbar artery/vein, and the lumbosacral plexus are exposed.
(9) Sacral (presacral) region dissection. After medial mobilization of the sigma-rectum, the peritoneum and the presacral fascia are incised medially to the right common iliac artery. Sacral lymph nodes, localized below the bifurcation of the abdominal aorta and inferior vena cava, in the triangle between the left and right common iliac vessels, are dissected [1,5,7,16–19] .

5. Anatomical Landmarks and Anatomical Variations, Related to Systematic PLND

Anatomy of the pelvic ureter. Identification of the ureter is a necessary step during PLND for two reasons: to avoid injury and to serve as a medial landmark during PLND. The ureter is divided into abdominal, pelvic, and intramural segments [20]. According to Luschka's law, the left ureter crosses the iliac artery 1.5 cm below the common iliac artery bifurcation, while the right ureter crosses the iliac artery 1.5 cm above the bifurcation. Therefore, the ureter enters the pelvic cavity by crossing the common iliac artery on the left side and the external iliac artery on the right [21,22]. Hence, the left ureter is located laterally to the internal iliac artery, whereas the right ureter is located medially to the right internal iliac artery. Surgeons should fully understand the anatomical relation between the ureter and iliac vessels. As the ureter enters the true pelvis, it runs caudally and medially to the ovarian vessels, reaching the bladder on the posterior leaf of the broad ligament [23].

6. Anatomical Variations of the Ureter, Related to PLND in Gynecologic Oncology (PLNDGO)

There is a multitude of ureteral anomalies, but the following three groups are related to PLND in gynecologic oncology (PLNDGO):

(A) Multiplication of ureter;
(B) Ureteral diverticulum;
(C) Unusual ureteral position—retro-iliac ureter.

(A) Ureteral multiplication is a longitudinal segmentation of the ureter into two or more tubes. Multiplication may be complete, incomplete, bilateral, and unilateral (Figure 6). Ureteral duplication is the most common type of multiplication and occurs more commonly in women than in men (4–5% of the population). Incomplete duplications tend to be unilateral and more common than complete, whereas complete duplications are often bilateral [24–26]. Unilateral ureteral duplication is more common on the right ureter than on the left (Figure 6) [25].

Surgical considerations. Ureteral multiplication increases the incidence of ureteral injury during PLND. Surgeons have to identify the course of the ureter as it crosses the pelvic brim and reaches the bladder. Injuries to the blood vessels of the ureters must be avoided, as duplicated ureters are usually contained within a single sheath and the associated blood supply could be interrupted [24,26]. Additionally, one of the duplicated ureters might be confused with vessel structures (artery most frequently), cut, and ligated during dissection. If intraoperatively, a suspicion of ureteral duplication arises, surgeons should observe the peristaltic activity of both ureters to differentiate this variation.

(B) Ureteral diverticulum is a rare urological congenital anomaly, classified into three subgroups: (1) abortive ureteral duplications, sharing the same embryogenesis as diverticula of the disordered ureteral budding; (2) true (congenital) diverticulum, characterized by the presence of all tissue layers of the normal ureter; and (3) false (required) diverticulum, which represents a mucosal herniation [27,28]. Ureteral diverticula are mainly asymptomatic, although urinary tract infection and ureteral stones causing obstruction could appear. Papin and Eisendrath proposed and illustrated urethral diverticulum classification, which can be used clinically for PLNDGO (Figure 6) [29].

Surgical considerations. Ureteral diverticulum may also be confused with vascular structures. Ampullary diverticulum could mimic retroperitoneal cysts. In cases of retroperitoneal tumors in the pelvic region during PLNDGO, the ureter must be identified.

(C) Retro-iliac ureter (RIU) is a rare urological condition in which the ureter passes deep to the iliac vessels. Generally, the condition is diagnosed intraoperatively. Despite being a congenital anomaly, for the majority of patients the RIU first manifests itself as flank pain and symptoms of ureteral obstruction in their second or third decade of life. Coexisting anomalies are common: vaginal atresia, lumbosacral agenesis, or malrotation of the kidney. Surgical treatment consists of a dissection of the ureter and its anterior repositioning with a subsequent reimplantation in the bladder wall.

RIU is thought to originate as a result of embryologic defects: a defect in the mesonephric ureteral migration during arterial development, persistence of the embryologic primitive ventral root of the umbilical artery, between the aorta and distal umbilical artery, traps the ureter dorsally. The third hypothesis involves abnormal development of the iliac vessels from the anterior branch of the umbilical artery instead of the dorsal [30–32]. RIU is commonly located behind the common iliac artery, but other previously reported types have been described as a retro-common iliac vein, retro-external iliac artery/vein, and retro-internal iliac artery [30–32]. A bilateral RIU is a rare but possible variation (Figure 6) [30].

Surgical considerations. Although a rare anatomic variation, the RIU is of high importance. First, it is hard to be identified if the surgeon is not familiar with such a variation of the ureter. Second, it could be injured during an iliac lymph nodes dissection, especially if it is mistaken for a vascular structure. Finally, the RIU could not be the dorsomedial border of the PLND, as it has different location. In such cases, the lateral aspect of the rectum is considered as the dorsomedial border.

Anatomical variations of the ureter—conclusion of surgical considerations.

Ureteral injuries (UIs) occur approximately in 0.5–1% of all pelvic operations with gynecological operations accounting for 75% of these. UIs are more common during radical hysterectomy procedures combine with PLND. The incidence of UIs during PLND is 10%. It is believed that ureteral variations lead to an increased risk of UIs. There is consensus that a multiplication of the ureter is an independent risk factor for UIs during PLND. Benedetti-Panici et al. reported a prospective study, which included 309 consecutive patients with cervical, endometrial, and ovarian cancer treated with systematic aortic and PLND. Ureteral duplication was observed in four (1.6%) cases. Authors concluded that knowledge of ureteral anomalies is important in gynecological surgery, as the risk of UIs is elevated. Duplicated UIs during gynecological procedures are reported in medical literature. UIs may occur at the level of the infundibulopelvic ligament and deep in the pelvis, below the level of ischial spine, where the ureter lies lateral to the peritoneum of uterosacral ligament. Anatomical variations of the ureter are often associated with other congenital anomalies. Patients with known congenital anomalies or different types of syndromes should undergo preoperative imaging for careful surgical planning [33–39]. Postoperatively, a routine cystoscopy should be performed to rule out UIs [26].

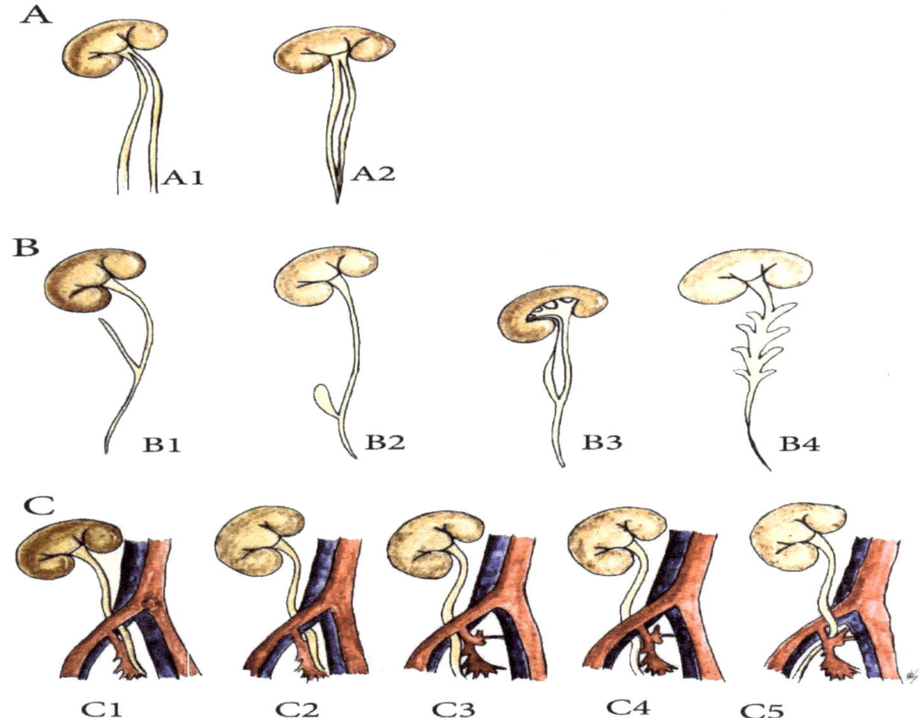

Figure 6. Anatomical variations of the ureter related to pelvic lymph node dissection in gynecologic oncology (PLNDGO). (**A**) Duplicated ureter. (**A1**) complete duplication. (**A2**) incomplete duplication. (**B**) Ureteral diverticulum (Adapted from Papin and Eisendrath [29]). (**B1**) simple diverticulum, (**B2**) ampullary diverticulum, (**B3**) diverticulum ending in fibrous prolongation, (**B4**) multiple diverticulus. (**C**) Retroiliac ureter. (**C1**) Behind the common iliac artery, (**C2**) behind the common iliac vein, (**C3**) behind the external iliac artery, (**C4**) behind the external iliac vein, (**C5**) behind the internal iliac artery.

7. Iliac Vessel Variations

7.1. Iliac Arteries

As PLND is mainly a vascular dissection procedure, the anatomical variations of iliac vessels should be respected in order to avoid unnecessary hemorrhage or transfusion [40].

7.2. Common Iliac Artery (CIA) Anatomy

The abdominal aorta bifurcates anterolaterally to the left side of the fourth lumbar vertebral body and divides into the left and right common iliac arteries. They further divide into the external and internal iliac arteries at the level of the sacroiliac joint. The right CIA (5 cm) is frequently longer than its left counterpart (4 cm) [41].

7.3. CIA Variations Related to PLNDGO

CIA variations are relatively rare. Although the true incidence of CIA variations is not known, the most frequent anomaly is an absence of the CIA. It tends to be unilateral, predominantly to the right, even though bilateral agenesis of CIA has also been described [42–44]. In one of the cases, described by Shetty et al., the abdominal aorta was directly branching into external and internal iliac arteries at the level of the 4th–5th lumbar vertebra [43]. Dabydeen et al. described a case of congenital absence of the right CIA. In their article, the proximal part of the right external iliac artery was absent. The distal part and the right common femoral artery was reconstituted from the right inferior epigastric artery, deep circumflex iliac artery, and the contralateral common femoral artery [44]. Llauger et al. presented a case of atresia of the right CIA, associated with a large aberrant and anomalous artery, connecting both hypogastric arteries within the pelvis [45].

Rusu et al. observed the anomalous origin of the iliolumbar artery, which has not been reported before. The authors dissected 15 human adult cadavers (30 pelvic halves). The CIA appeared trifurcated due to a higher origin of the iliolumbar artery (originating from the CIA bifurcation) in 2.5% of cadavers [46]. In 8.75% of specimens, the iliolumbar artery originated from the CIA [46].

Surgical considerations. In cases of absent CIA, surgeons should be aware of collateral, anomalous, or aberrant arteries. Collateral arteries serve as an alternative blood source, compensating the absent CIA; they might be of atypical origin from the iliac system. Iliolumbar arteries, originating from the CIA or CIA bifurcation, may be damaged during a middle common iliac lymph nodes dissection of the lumbosacral fossa.

8. External/Internal Iliac Artery Anatomy

The external iliac artery (EIA) is a direct continuation of the CIA. It runs downwards, forwards in the iliac fossa, and reaches the lacuna vasorum under the inguinal ligament. The EIA has two branches: inferior epigastric artery and deep circumflex iliac artery. After crossing the mid-inguinal point, it continues as a common femoral artery. The internal iliac artery (IIA) arises from the CIA anterior to the sacroiliac joint. In the majority of the cases, the artery originates at the level of the L5–S1 intervertebral disc. The IIA descends posteriorly towards the superior border of the greater sciatic foramen where it divides into two branches: anterior and posterior [42,47,48]. The anterior and posterior branches of the IIA are illustrated in Figure 7.

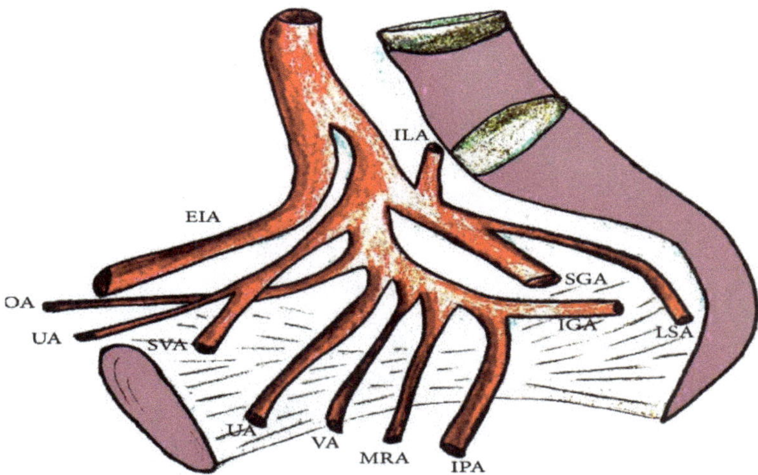

Figure 7. The internal iliac artery (IIA) branches. OA—obturator artery, UA—obliterated umbilical artery, SVA—superior vesical artery, UA—uterine artery, VA—vaginal artery, MRA—middle rectal artery, IPA—internal pudendal artery, IGA—inferior gluteal artery, LSA—lateral sacral artery, SGA—superior gluteal artery, ILA—iliolumbar artery.

9. EIA Variations Related to PLNDGO

Variations of the EIA are grouped into four categories [48,49]:

(A) Hypoplasia, agenesis of the EIA;
(B) Anomalous origin and position of the EIA;
(C) Morphological variations (variations in the shape of the EIA);
(D) Variations in the branching pattern of the EIA.

(A) Kawashima et al. reported a rare case of hypoplastic right EIA, which continued into the normal femoral artery by anastomoses formed with the enlarged obturator and deep circumflex iliac arteries [50]. Okamoto et al. reported a case of the CIA entering into the pelvic cavity without branching to the EIA. The artery passed behind the first sacral nerve and gave rise to each branch of the IIA in the pelvis [51]. In cases of EIA hypoplasia, an ipsilateral persistent sciatic artery may also be observed [49].

Surgical considerations. Cases of EIA hypoplasia and agenesis are of clinical significance due to the possible presence of a collateral pathway. PLND should be meticulous in order to preserve collateral and anomalous vessels.

(B) The EIA may directly originate from the abdominal aorta. The origin of the EIA could be posterior in the internal iliac fossa. The position of the EIA may differ to the external iliac vein (EIV) position. The EIA might be located superficially or medially to the EIV [48,49].

Surgical considerations. During an external iliac lymph nodes dissection, surgeons should be familiar with the EIA position differences to prevent iliac vessel injury. Safi et al. reported a case of EIA laparoscopic injury due to an anomalous position of the artery. The EIA was elongated and located posterior and medial to the EIV. They repaired the artery using a two-needle single-knot technique and continued the lymphadenectomy. Authors concluded that during PLND, both external iliac vessels should be exposed and visible to the surgeon, as the anatomical variations of the EIA increased the risk of injury [52].

(C) Morphological variations of the EIA are grossly classified in five groups: looped, tortuous, curved, twisted, and S-shaped EIA. Nayak et al. presented a cadaveric study of 48 hemipelvises. They observed morphological variations in nine (19%) of the hemipelvises [48]. Looped and tortuous EIA have been reported in other case studies [53,54]. We also observed morphological variations of EIA (Figure 8).

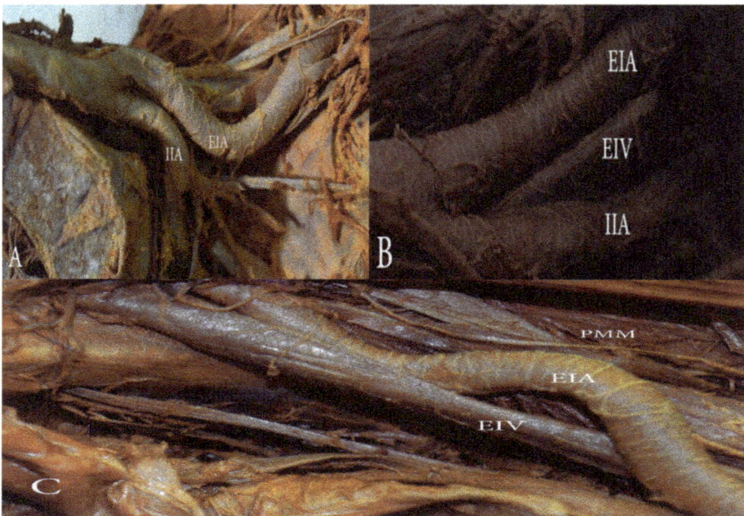

Figure 8. Morphological variations of the external iliac artery (EIA). (**A**) The EIA with an inward loop in the left hemipelvis. (**B**) The EIA with a gentle inward loop in the left hemipelvis. (C) 'S' shaped EIA in the right hemipelvis. EIA—external iliac artery, EIV—external iliac vein, IIA—internal iliac artery, PMM—psoas major muscle.

Surgical considerations. Knowledge of the EIA morphological variations is essential during an external iliac lymph nodes dissection. Particular attention should be paid during dissection of the lateral, middle, and medial external iliac lymph nodes. The risk of iatrogenic EIA lesions may increase as a result of existing morphological differences.

(D) Variability in the origin of the obturator artery (OA) from the EIA has also been reported in medical literature [55,56]. If the OA originates from the external iliac system, it crosses the superior pubic ramus and the external iliac vein vertically [55]. The OA can also arise as a common trunk with the inferior epigastric artery and the deep circumflex iliac artery (Figure 9) [55,56]. Nayak et al. reported duplication of the deep circumflex iliac artery and an additional large muscular branch, arising and descending laterally from the EIA [48]. The medial circumflex femoral artery and the profunda femoral artery could also arise from the EIA [42].

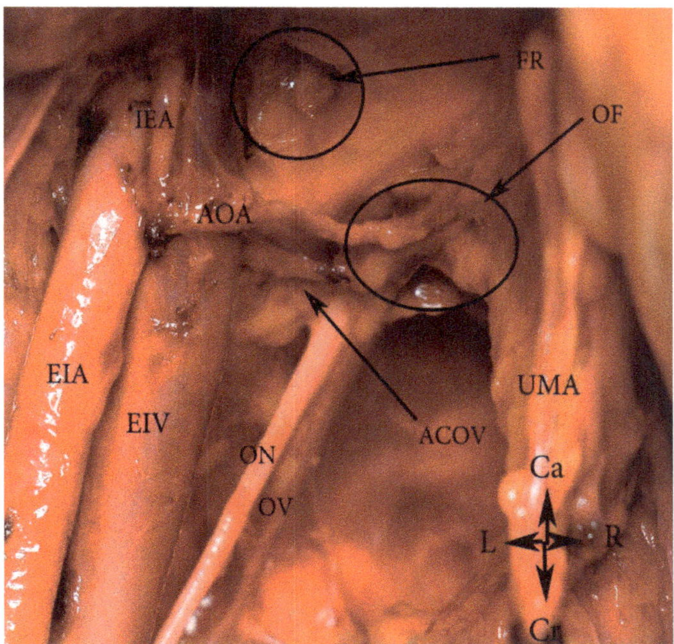

Figure 9. An aberrant obturator artery arising as a common trunk with the inferior epigastric artery (open surgery). EIA—external iliac artery, EIV—external iliac vein, AOA—aberrant obturator artery, ACOV—accessory obturator vein, ON—obturator nerve, OV—obturator vein, IEA—inferior epigastric artery, UMA—umbilical artery, FR—the deep femoral ring, OF—obturator foramen. Ca—caudal, Cr—cranial, R—right, L—left.

Surgical considerations. An OA, arising from the EIA, is not a rare variation. The majority of OA origin variations are located at the distal part of the EIA.

10. IIA Variations, Related to PLNDGO

Most of the IIA variations are related to its branching pattern. Different authors have suggested various concepts for classification of the ending branches of the IIA. Most researchers and clinicians refer to the Adachi classification, which has been the standard for many years. Adachi classified the branching of the IIA using its four large parietal branches: the umbilical, superior gluteal, inferior gluteal, and the internal pudendal arteries (Figure 10) [42,57–59].

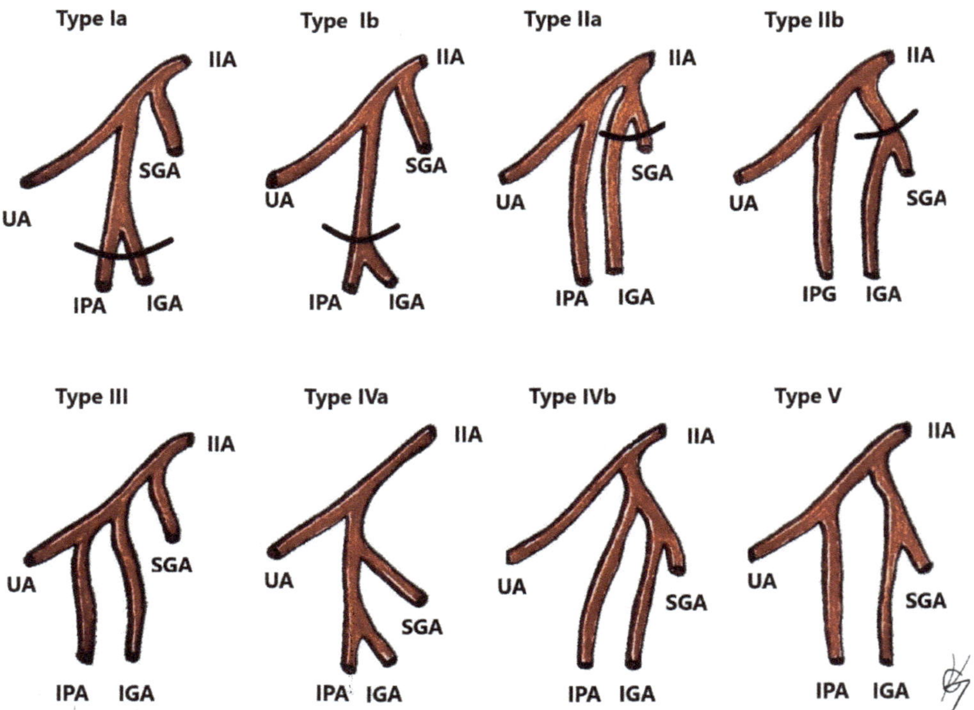

Figure 10. Classification of IIA variations. Adopted from Adachi [58]. Type I—The superior gluteal artery (SGA) arises separately from internal iliac artery, while the inferior gluteal (IGA) and internal pudendal vessel (IPA) share a common trunk. Type Ia—the bifurcation of IGA and IPA occurs within the pelvis. Type Ib—the bifurcation occurs below the pelvis. Type II—The internal pudendal artery arises separately from the IIA, while the superior gluteal artery shares a trunk with the inferior gluteal artery. Type IIa—the bifurcation of SGA and IGA occurs within the pelvis. Type IIb—the bifurcation occurs below the pelvis. Type III—SGA, IGA, and IPA arise separately from the internal iliac artery, and the internal pudendal artery is the last branch. Type IV—SGA, IGA, and IPA share a common trunk. Type IVA—the SGA is the first vessel from the common trunk, before bifurcating into the other two branches—SGA and IGA. Type IVB—the IPA is the first from the common trunk, which then divides into SGA and IGA. Type V—The IGA has a separate origin from the IIA, while the SGA and IGA share a common trunk.

Surgical consideration. As shown in Figure 10, the SGA is a variable artery. Surgeons should be aware of the SGA variations to safely perform gluteal lymph node group dissection.

11. Uterine and Obturator Artery Variations Related to PLNDGO

Uterine artery. In the majority of cases, the uterine artery (UA) arises from the anterior branch of the IIA. Different origins of the UA have been reported: from the umbilical artery as a separate branch and via a common trunk (Figure 11); from the inferior gluteal artery as a separate branch; as a bifurcation from the inferior gluteal; and as a trifurcation with the superior and the inferior gluteal arteries [60].

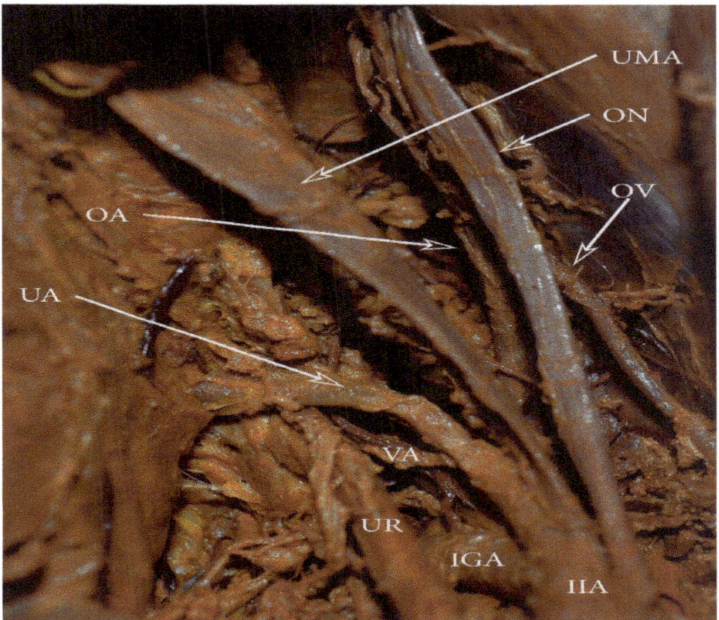

Figure 11. Uterine artery and umbilical artery, arising in a common trunk. VA arising from the uterine artery. OA—obturator artery, OV—obturator vein, ON—obturator nerve, UMA—obliterated umbilical artery, IIA—internal iliac artery, IGA—inferior gluteal artery, UR—ureter, VA—vaginal artery.

Surgical considerations. The risk of iatrogenic injury to the UA is increased when the artery arises from the umbilical artery, as the UA crosses the operative field. If the PLND is performed before hysterectomy, surgeons should first identify the UA origin.

Obturator artery. The OA, branching from the EIA, has been discussed above. Herein, most of the OA anomalies will be analyzed.

In most cases, the OA originates from the anterior trunk of the IIA [55]. The OA runs anteroinferiorly and lies longitudinally to the obturator foramen on the medial part of the obturator internal muscle. It gives several branches within the pelvis, before entering the obturator foramen. They are classified as iliac, vesical, and pubic branches. The OA is located medially to the ureter, cranially to the obturator vein, and caudally to the obturator nerve [55,61]. The OA has the greatest variation frequency among the IIA branches [62]. Furthermore, the OA may arise from the EIA, the femoral artery, the deep circumflex iliac artery, the posterior branch of IIA, or from the inferior epigastric artery (Figure 12) [42,55,63]. In 20% to 34% of the cases, the pubic branch of the inferior epigastric artery replaces the OA. In such cases, the OA passes posterior to the lacunar ligament and courses into the superior pubic ramus vertically to enter the obturator foramen [55]. An OA arising from the EIA or its branches is classified as an aberrant obturator artery. Two obturator arteries could be observed during dissection. An additional OA with a different origin or path through the obturator fossa is classified as an accessory obturator artery.

Figure 12. Obturator artery arising from the posterior branch of the IIA. U—uterus, EIA—external iliac artery, EIV—external iliac vein, IIA—internal iliac artery, SGA—superior gluteal artery, LSA—lateral sacral artery, IPA—internal pudendal artery, OA—obturator artery, ON—obturator nerve, SUA—severed uterine artery.

Surgical consideration. Surgeons should be aware of the possible presence of accessory or aberrant obturator arteries during external and internal iliac lymph node groups dissection. The presence of normal OA does not exclude the existence of an accessory OA. The inferior epigastric artery is located below the deep circumflex iliac vein. Any injury of the OA arising from the inferior epigastric artery should be avoided, as the deep circumflex iliac vein is the ventral border of the dissection.

Anatomical variations of iliac arteries—conclusion of surgical considerations.

Arterial injuries during PLNDGO are less common than venous ones. There are very few studies describing arterial damage during PLNDGO. Bae et al. presented a retrospective review study. Authors observed four (1.3%) cases of major vascular injuries during 225 laparoscopic lymph node dissections. One of the injuries was to the EIA. Ricciardi et al. and Ishikawa et al. also reported damage to the EIA in a course of PLND. These studies showed that the EIA is more likely to be injured during PLND than the other iliac arteries. Gyimadu et al. concluded that the occurrence of vascular injuries during the course of PLNDGO may be due to variations in the anatomy of great retroperitoneal vessels. In medical literature the frequency of such abnormalities varies between 5.6 and 23.0% [64–68].

12. Iliac Veins

12.1. Common Iliac Vein Anatomy (CIV)

Common iliac veins (CIVs) drain into the inferior vena cava. They are a direct continuation of the external iliac veins and are formed by the junction of the external and internal iliac venous system, anterior to the sacroiliac joints. The left CIV is longer than the right one. The left CIV is located first medially, then posteriorly to the left EIA, whereas the right CIV is posterior and then lateral to the right EIA. The iliolumbar, the ascending lumbar vein, and the lateral sacral vein drain into the CIV. In most cases, the median sacral vein drains into the left CIV [41,42,47].

12.2. CIV Variations Related to the PLNDGO

Surgical considerations. All types of CIV variations are related to possible venous injuries (Figure 13) [35,69–74]. Iatrogenic damage could occur during dissection of all five types of common iliac lymph node groups. Kose et al. reported a study of 229 patients who underwent paraaortic and PLND. Authors observed major retroperitoneal vessel variations in thirty nine (17%) patients. Great vessel injury was present in nineteen (8.3%) patients. CIV variations were found in two patients. One of the patients had a venous annulus of the right CIV, surrounding the right CIA. The other patient had a duplicated left CIV (B2 from the CIV classification), which was injured during dissection. Authors concluded that each patient's vascular anatomy must be assessed individually to avoid injuries during scheduled operations [75].

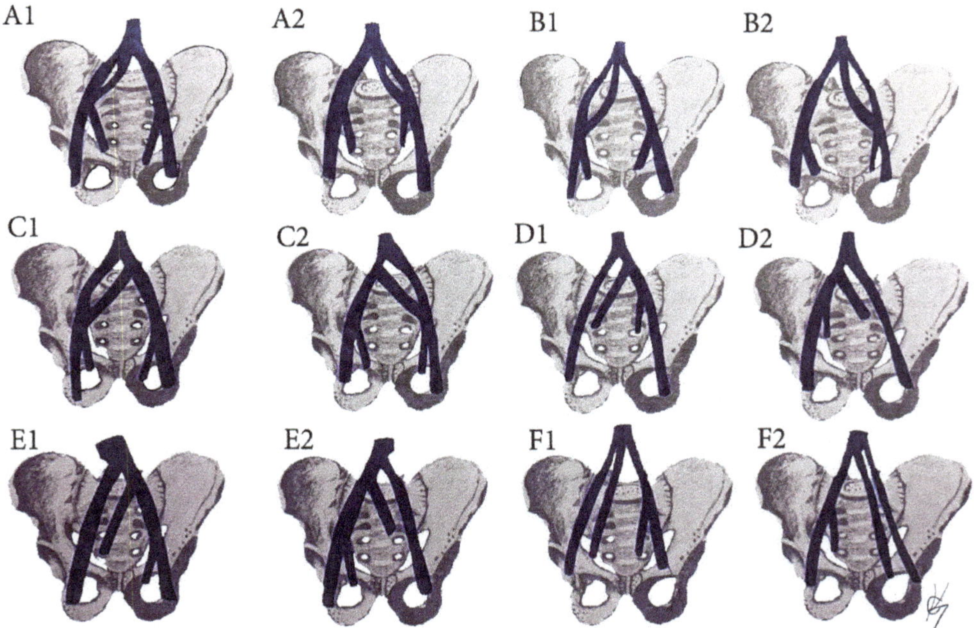

Figure 13. Common iliac vein variations. (**A**) Incomplete duplication of the CIV; (**B**) complete duplication of the CIV; (**C**) lateral duplicated branch drains into the IVC, the medial drain into the CIV. (**D**) Absent CIV, external and internal iliac veins drain to the contralateral CIV; (**E**) absent CIV, the EIV drains into the IVC, the IIV drains into the contralateral CIV; (**F**) Absent CIV, the external and internal veins drain into IVC. Inferior vena cava (IVC), Common iliac vein (CIV), external iliac vein (EIV), internal iliac vein (IIV). (**A1,B1,C1,D1,E1,F1**) are related to right hemipelvises variations, whereas (**A2,B2,C2,D2,E2,F2**) are related to left hemipelvises variations.

12.3. CIV Tributaries Variations Related to PLNDGO

Iliolumbar vein (ILV) and ascending lumbar vein (ALV) anatomy. The ILV drains the venous blood from the iliac fossa, the iliac, and psoas muscles and terminates in the CIV. It is considered the segmental equivalent of the fifth lumbar vertebra [76–81]. The ALV participates in an anastomotic venous system between the inferior vena cava and the superior vena cava. The lower end of the ALV enters the cephalic border of the CIV. Upwards, the ALV receives the lumbar veins and terminates by joining the subcostal vein to form the azygos vein on the right and the hemiazygos on the left. ILV anastomoses with ALV, deep circumflex iliac vein and lateral sacral vein [76–81].

ILV and ALV variations related to PLNDGO. Particular attention should be given to the ILV and AVL as high percentage of drainage variations is documented. Furthermore, our literature survey revealed that the clinical significance of these veins is rarely mentioned in gynecologic oncology practice. In PLNDGO, we observed a high percentage of drainage variations of the ILV and ALV. In medical literature, there is a controversy as to the anatomy and the nomenclature of the ILV and ALV. Terms such as "lateral lumbosacral veins", "inferior lumbar", and "superior iliac" veins have been used to define ILV and ALV [76–81]. Moreover, different drainage patterns of ILV and ALV have been reported for both veins—ILV/ALV draining separately into CIV; ILV/ALV draining into the CIV as a common trunk, ILV draining into the external/internal iliac venous system. Lolis et al. reported a detailed description of the surgical anatomy and draining patterns of the ILV, based on a significantly great number of specimens [77]. They proposed and illustrated a detail classification separated into two types. In Type I (54%), ILV drainage patterns differed, whereas the ALV had the same pattern on both sides. In Type II, the ALV differed in pattern from one side to the other (46%). Authors observed high percentage of drainage variations in ILV 91% compare to ALV 34% [77]. Numerously drainage variations of ILV and ALV have been reported, but in Figure 14 are illustrated the most important during PLNDGO. The ILV and ALV drainage variations during PLND in our practice are shown in Figure 15.

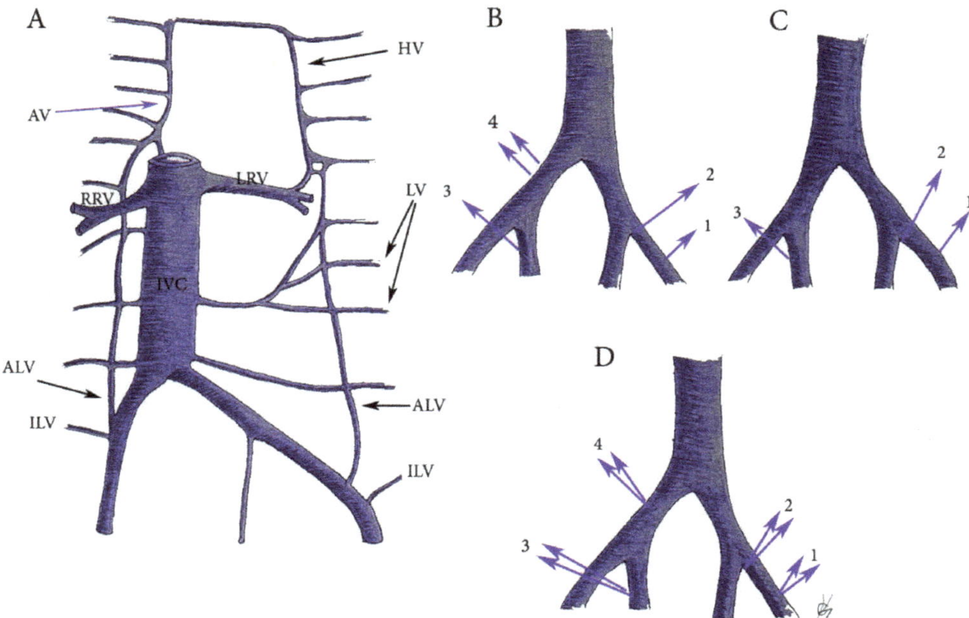

Figure 14. ILV and ALV anatomy and variations. (**A**) ILV and ALV anatomy. HV—hemiazygos vein, LV—lumbar veins, ALV—ascending lumbar vein, ILV—iliolumbar vein, IVC—inferior vena cava, LRV—left renal vein, RRV—right renal vein, AV—azygos vein. (**B**) ILV variations. 1—drains into EIV, 2—drains into the confluence of the CIV, 3—drains into the IIV, 4—two ILVs drains into the CIV. (**C**) ALV variations. 1—drains into the EIV, 2—drains into the confluence of the CIV, 3—drains into the IIV. (**D**) Common trunks between ALV and ILV. 1—drains into the EIV, 2—drains into the confluence of CIV, 3—drains into the IIV, 4—drains into the CIV.

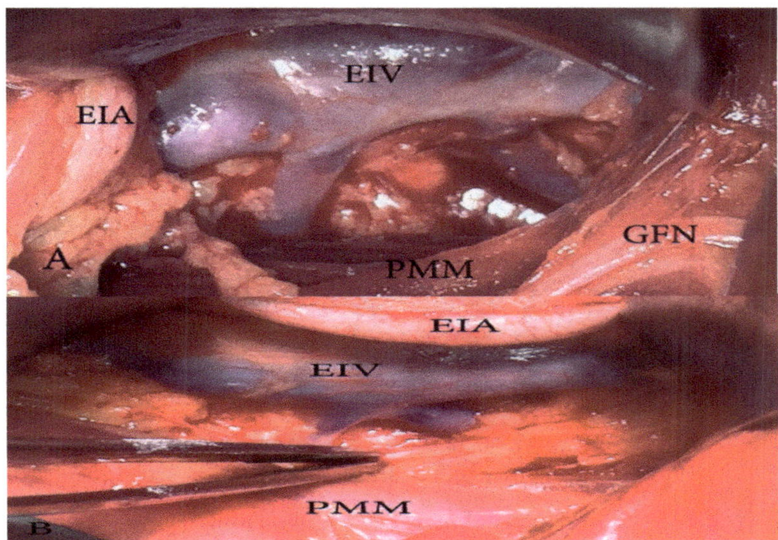

Figure 15. ILV or ALV drain into the EIV (open surgery right pelvic sidewall). We can only speculate if these veins are ILV, ALV, or both. (**A**) Two separate veins drain into the EIV. The EIA is retracted medially. (**B**) Two veins drain into the EIV via common trunk.

Surgical considerations. Knowledge of the surgical anatomy of ILV and ALV may prevent venous damage such as tears and avulsion of these veins during PLND. Injury of ILV and ALV could occur in the course of external and internal iliac lymph nodes dissection. Special attention should be paid during middle common iliac (located in the lumbosacral fossa) and lateral external iliac lymph nodes dissection. Panici et al. stated that during lateral common iliac lymph nodes dissection, the presence of iliolumbal veins could be hazardous as several iliolumbar veins could drain into the CIV. Authors concluded that the CIV should be handled very gently, and dissection must be blunt and delicate [17].

13. Median Sacral Vein (MSV) and Lateral Sacral Veins (LSVs) Anatomy and Variations Related to PLNDGO

The median sacral vein (MSV) runs anterior and in the midline of the sacrum and the coccyx. It commonly drains into the left CIV. The lateral sacral veins lie on the periosteum of the sacrum and typically connect the epidural plexus with the internal iliac veins. The MSV might drain into the left internal iliac vein or the common iliac junction. Anastomoses between the lateral and median sacral veins form the presacral venous plexus. Cardinot et al. reported a case of both internal iliac veins, which formed a common trunk with a short and an average course, receiving the middle sacral vein's drainage and flowing into the left external iliac vein. The lateral sacral veins (LSVs) might drain into the CIV and external iliac vein [41,47,72,82].

Surgical considerations. The MSV and the LSVs have to be preserved in cases of presacral and lateral sacral lymph node group dissection. Surgeons should be aware of different drainage patterns and venous plexus existence between the two veins.

External iliac vein (EIV) anatomy. The EIV is the continuation of the femoral vein. The inferior epigastric vein, the deep circumflex iliac vein, and the pubic branch drain into the EIV. The vein is located medially to the ipsilateral homonymous artery.

14. EIV Variations Related to PLNDGO

Anatomical variations of the EIV are less common than the CIV and internal iliac vein. The EIV might double, be absent, or be located lateral to the homonymous artery [82–84].

Hayashi et al. reported a case of an additional right EIV, which originated 45 mm inferior to the iliocaval junction and ran ventrally to the EIA to surround it with a right EIV. The right CIV was absent [84]. Djedovic and Putz observed a case of a venous annulus of the left external iliac vein. The medial and the lateral branch of the left EIV surrounded the left EIA. Moreover, a communication branch, between the lateral and the medial branches of the EIV, was identified. It was located below the left EIA [83].

Surgical considerations. Lateral, additional, double EIV, or venous annulus might be injured during a dissection of external iliac lymph nodes. Lateral, middle, and median external iliac lymph nodes are at great risk of iatrogenic damage. EIV injuries during PLNDGO have been reported in medical literature [85,86]. Roda et al. reported two cases of EIV injury among 327 pelvic lymphadenectomies for gynecological malignancies [86]. Kose et al. reported a case where damage to the EIA was due to supernumerary renal artery and vein, which distorted the normal anatomy [75].

15. EIV Tributaries Variations Related to PLNDOG

Deep circumflex iliac vein (DCIV) anatomy and variations related to PLNDG. The deep circumflex iliac vein (DCIV) runs over the EIA and above the inguinal ligament, it drains in the EIV. It is known that the draining pattern of the DCIV is variable. Ghassemi et al. observed that the DCIV ran over (82.5%) and under (17.5%) the EIA. Their study included 216 hemipelvises—78 cadavers and 60 clinical cases [87].

Surgical considerations. The DCIV under the EIA is less likely to be identified. Such an instance may lead to an expanded pelvic lymph node dissection (PLND). As mentioned above, the DCIV is the ventral border of PLND. The DCIV under the EIA is visualized between the EIA and EIV or by medial traction of the EIV (Figure 16).

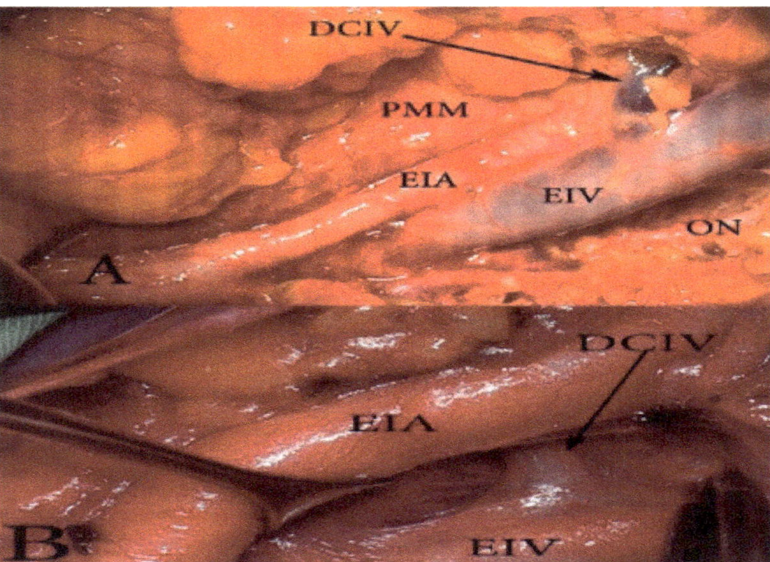

Figure 16. The deep circumflex iliac vein (DCIV) normal anatomy (**A**) and variation (**B**) (open surgery, left pelvic sidewall). (**A**) The DCIV runs over the EIA and drains into the EIV. (**B**) The DCIV passes under the EIA and drains into the EIV.

16. Internal Iliac Vein (IIV) Anatomy

IIV follows its named arterial counterpart and ascends posteromedial to the IIA. IIV drains towards the ipsilateral EIV. The IIV tributaries are the superior/inferior gluteal, obturator, internal pudendal, lateral sacral, middle rectal, superior/inferior vesical, uterine, and vaginal veins. The retroperitoneal venous system is derived from the modification of three parallel primary venous networks in the embryo between the sixth and tenth weeks of gestation—the subcardinal, the postcardinal, and the supracardinal veins [41,42,88,89].

17. IIV Variations Related to PLNDGO

The multiple anomalies in the hypogastric venous drainage system represent posterior cardinal vein maldevelopment, as the posterior cardinal veins form the iliac bifurcation and iliac veins [75]. IIV variations have not been studied as thoroughly as IIA variations [79,89–92]. Despite the various classifications describing the diverse variations of the IIV, there is no established standard classification [79,89–94]. Shin et al. have created an impressive, comprehensive, and generally reliable classification of iliac vessel variations based on 2488 patients using multidetector computed tomography [90]. However, other types of IIV variations exist, which have not been mentioned in Shin's classification. Therefore, to clarify most of IIV variations, a modification of Shin's classification was developed based on previous findings (Figures 17 and 18) [42,73,79,89–94].

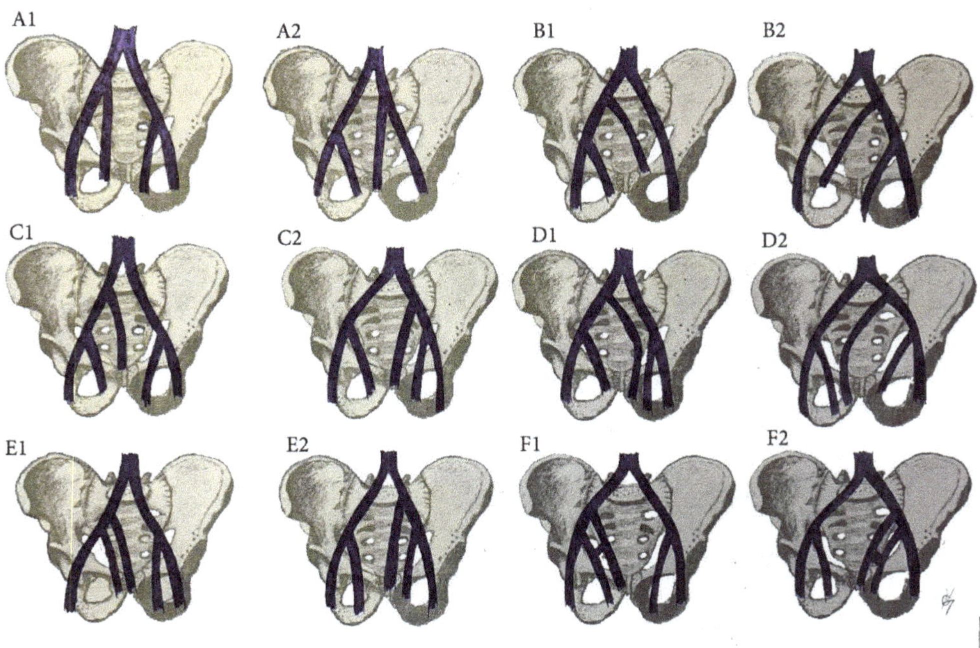

Figure 17. IIV variations. (**A1**) high joining of the IIV to the ipsilateral EIV. (**B1**) Joining of the IIV to the contralateral CIV. (**C1**) Separated trunk of the IIV drains into the ipsilateral CIV. (**D1**) separated trunk of the IIV drains into the contralateral CIV. (**E1**) Duplication of the IIV. (**F1**) Duplication of the IIV with a venous connection between them. Variations 1 are related to right pelvic sidewall, whereas variations 2 are related to the left pelvic sidewall.

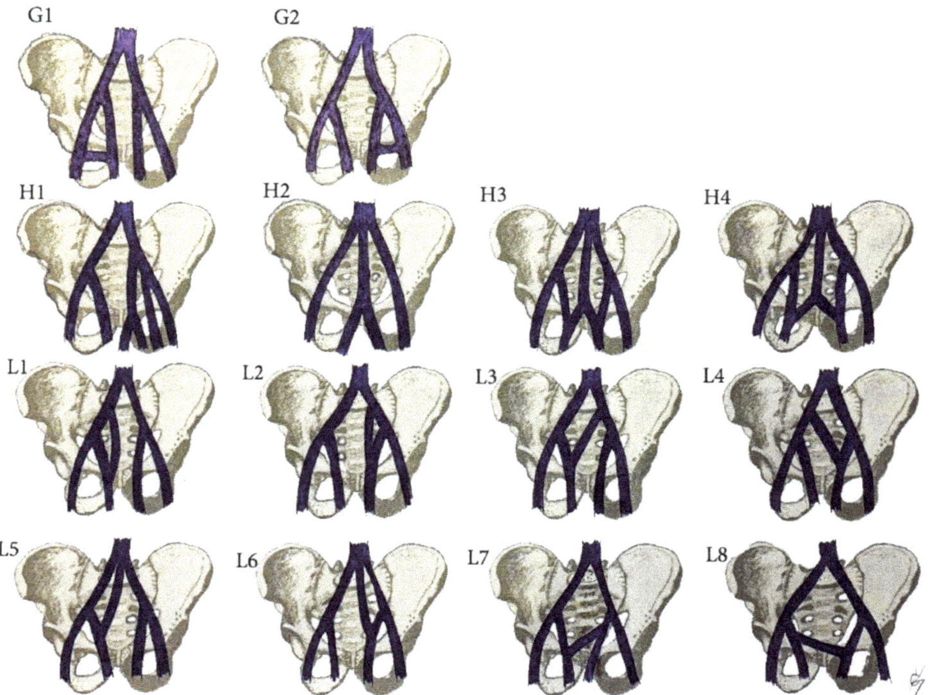

Figure 18. IIA variations. (**G1**) Communication vein between the IIV and the EIV. (**H1**) Separated trunk of bilateral internal iliac veins connecting with each other before draining into the left CIV. (**H2**) The internal iliac veins form a common trunk, that drains into the inferior vena cava. (**H3**) The internal iliac veins form a common trunk, which drains into the inferior vena cava, communication vein between the IIVs and ipsilateral EIV. (**H4**) Both IIVs are joined by a connecting vein that drains into the IVC. (**L**) Communication veins. (**L1,L2**) Communication vein between the IIV and ipsilateral CIV. (**L3,L4**) communication vein between the IIV and contralateral CIV. (**L5,L6**) Both internal iliac veins are joined with a communication vein, which drains into the inferior vena cava. (**L7,L8**) Communicating vein between both IIVs. Variations (**G1,L1,L3,L5,L7**) are related to right pelvic sidewall, whereas variations (**G2,L2,L4,L6,L8**) are related to left pelvic sidewall.

Surgical considerations. As shown in Figures 17 and 18, the prevalence of IIV variations is high. These variation veins may cause problems of unexpected hemorrhage during dissection of all lymph node groups. Therefore, it is crucial to recognize the presence of these variations, which are often encountered intraoperatively. Gyimadu et al. reported three (8.1%) cases of left duplicated IIV injuries (E2 from the IIV classification) among 37 patients with anatomical vessel variations. All of the patients underwent PLND and paraaortic lymphadenectomy for gynecological malignancies. Authors concluded that anatomical vessel variations are not uncommon and may increase the risk of vascular complications during PLND [67]. Panici et al. described the frequency of retroperitoneal variations among 309 consecutive patients with cervical, endometrial, and ovarian cancer treated with systematic aortic and PLND. Authors observed three (1.3%) cases of right IIV draining into the left CIV (B2 from the IIV classification). There were no cases of intraoperative injury to these veins [37].

Anatomical variations of iliac veins—conclusion of surgical considerations.

Iliac vein variations and injuries are more common than arterial ones. The three-dimensional (3D) models are reconstructed on the basis of multi-detector computed tomography. The surgeons observe the reconstructed 3D models and identify all of the anatomical structures before surgery. The 3D models of the pelvic vessels may help avoid injury to anatomical vessel variations during PLNDGO by providing information on individual anatomical features before gynecological procedures [95].

18. Corona Mortis, Aberrant and Accessory Obturator Veins Related to PLNDGO

Corona mortis (CMOR) is defined as any vessel anastomoses between the external iliac and obturator vessels, excluding aberrant and accessory obturator vessels. These originate from the external iliac or the inferior epigastric vessels and pierce the obturator membrane, not participating in anastomoses [55,96]. The CMOR could be arterial, venous anastomoses, or both. The frequency of venous CMOR is higher than the arterial. The CMOR is located behind the superior pubic ramus and on the posterior aspect of the lacunar ligament [55,96–98].

Aberrant and accessory obturator veins could arise from the EIV and its tributaries. The definition of an aberrant obturator vein is a vein that drains into the EIV system. There is no other obturator vein (Figure 19).

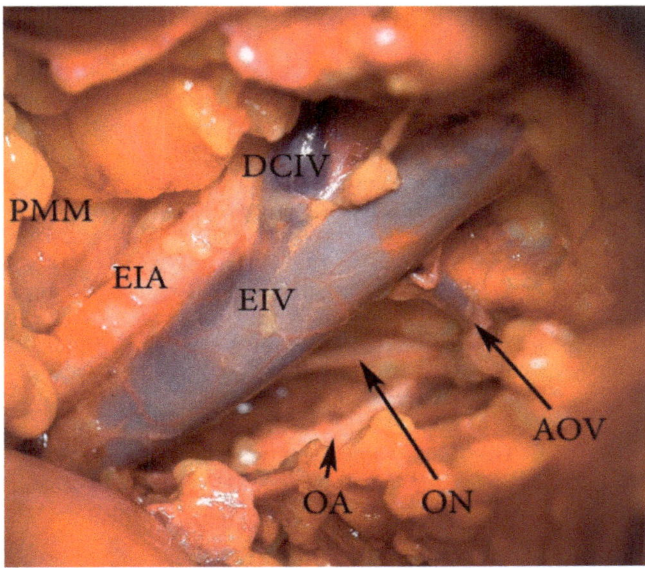

Figure 19. An aberrant obturator vein (left pelvic sidewall). EIA—external iliac artery, EIV—external iliac vein, PMM—psoas major muscle, AOV—aberrant obturator vein, ON—obturator nerve, OA—obturator artery, DCIV—deep circumflex iliac vein.

An accessory obturator vein is an extra obturator vein, draining into the EIV system, in addition to the normal counterpart [47,96–98].

Surgical consideration. Damaging the CMOR, aberrant and accessory obturator vessels could occur throughout medial external iliac and obturator lymph node group dissection. Aberrant or accessory obturator veins have vertical direction through the obturator canal. Injuring of these vessels is more troublesome than CMOR injury, as obturator nerve and artery are located nearby and should be preserved. Lee et al. reported two (10.5%) cases of aberrant obturator veins injury during 19 PLND for gynecological malignancies [99]. Selcuk et al. reported four (4.1%) cases of CMOR injuries among 209 patients who underwent PLNDGO [100].

19. Nerves Anatomy

19.1. Obturator Nerve (ON) Anatomy

The obturator nerve (ON) arises from the ventral roots of the second, third, and fourth lumbar nerves. It descends through the fibers of the psoas major muscle and emerges from its medial border. The ON crosses the sacroiliac joint behind the CIA, lateral to the internal iliac vessels travels along the lateral wall of the lesser pelvis and enters the obturator foramen. The ON is located cranial to the OA and OV [41,101–104].

19.2. ON Variations Related to PLNDGO

An accessory obturator nerve (AON) could arise from the anterior divisions of L2–L3, L3 only, L3–L4, from the ON, and from the femoral nerve. The AON is located medially to the femoral nerve and laterally to the ON. The nerve lies on the medial border of the psoas major muscle, but instead of piercing the obturator foramen, it passes over the superior pubic ramus. It runs behind pectineus and divides into three branches, which are also variable. The incidence of AON in the human population varies from 10% to 30%. Studies did not find differences of AON presence between genders [41,101–104].

Surgical considerations. Compression and subsequent neuropathy may occur as a result of damage to the AON [103]. Such an injury is possible during a dissection of the lateral external iliac, obturator, lateral, and middle common iliac lymph nodes.

19.3. Genitofemoral Nerve (GFN) Anatomy

The origin of the genitofemoral nerve (GFN) is from the ventral rami of L1 and L2 of lumbar plexus. It penetrates the psoas major muscle and runs cranially along the anterior aspect of the muscle, beneath the transversalis fascia and the peritoneum. In most cases, the GFN bifurcates into its both branches midway along the anterior surface of the psoas major. The genital branch follows the inguinal ligament and ends in the skin of mons pubis and labium majus. The femoral branch leaves the pelvis by passing through the femoral sheath lateral to the femoral artery and supplies the skin of the proximal anterior thigh [41,102,105–107].

19.4. The GFN Variations Related to PLNDGO

The GFN exhibits a large number of origin variations—T12-L1, L2-L3, L1, L2, and L3. Unilateral absence of the GFN has been reported. In such cases, the ilioinguinal nerve replaces the genital branch and the anterior femoral nerve or lateral cutaneous replaces the femoral branch. The genital or femoral branches of the nerve may arise separately [101,105–107]. Paul and Shastri observed the GFN in 60 hemipelvises. They reported for early division of the nerve into genital and femoral branches at its formation in 13.3% of hemipelvises or in the middle of its course, after emerging from psoas major in 3.3% of specimens [106]. Another study, reported that the most common variation of the GFN was splitting of the nerve into genital and femoral branches within the substance of the psoas muscle [102]. Injury to the GFN may cause entrapment neuropathy [106].

Surgical considerations. As the GFN is the lateral border of PLND, it should be identified on the psoas major muscle prior to PLND. Early division of the GFN into genital and femoral branches means that two nerve fibers would be identified on the psoas major muscle—genital and femoral. If the two nerve fibers are recognized on the psoas major muscle, they should be preserved to prevent neuropathy.

Anatomical variations of the GFN and the ON—conclusion of surgical considerations.

Cardosi reported a study of 1210 patients, who underwent major pelvic surgeries for gynecological malignancies. Twenty-three patients had postoperative neuropathies. The incidence of obturator nerve injury (39% of all neuropathies) was higher than for other nerve lesions. Genitofemoral neuropathy was identified in four (17.3% of all neuropathies) women who underwent PLND. The frequency of injury of variant ON and GFN during PLNDGO is uncertain, but it is believed to be higher than those with normal anatomy [108].

There are several strengths of the present article. First, such a comprehensive review of the topic has never been made. Second, despite the multitude of articles describing PLNDGO, authors did not mention differences in morphology of the EIA. Very few anatomical articles reported morphological differences of the EIA [48,53,54]. Third, the different drainage patterns of ILV and ALV have never been discussed in gynecologic oncology. An article presented by Cibula and Rustum illustrated the ILV draining into the EIV and the CIV [5]. Panici et al. discussed the importance of ILV draining into the CIV during PLNDGO [17]. In both articles, it is not mentioned that the ALV could also drain into the CIV, EIV, or IIV. These articles described the ILV as the only vein draining into the EIV or the CIV. Moreover, the ILV and the ALV may drain into the iliac venous system by sharing a common trunk.

A potential limitation of the present article is that some of the anatomical variations are rare and there is limited data about the actual incidence of complications during PLNDGO. A possible explanation about the limited data could be that injuries to variant anatomical structures are managed during surgery. Furthermore, injuries with fatal outcome are less likely to be reported. We encourage surgeons to share their experience with injuries to variant anatomical structures during PLNDGO in order to estimate the actual incidence of complications.

20. Conclusions

A wide variety of anatomical variations among pelvic structures (ureters, vessels, and nerves) could cause severe and potentially lethal complications during surgery. The majority of the anatomical variations are discovered intraoperatively. Therefore, a detailed knowledge of the anatomy and anatomical variations is essential in order to prevent serious damage to vital structures during pelvic operations. The present article aims to expand the limited knowledge about anatomical variations in the pelvis. An association between variations of the most important pelvic structures and PLND is conducted for the first time. We hope that the detailed review of the anatomical variations will decrease patient morbidity and mortality. Furthermore, accurate description and analysis of the majority of pelvic anatomical variations may impact not only gynecological surgery, but also spinal surgery, urology, and orthopedics.

Author Contributions: Conceptualization, S.K. and Y.K.; methodology, S.K., A.Y., and Y.I.; formal analysis, S.S., Y.I., and S.K.; investigation, S.K. and D.D.; resources, D.D., Y.I., and S.K.; data curation, S.S.; writing—original and draft preparation, S.K.; writing—review and editing, S.S., N.D., and Y.I.; visualization, D.D. and N.D.; supervision, A.Y. All authors have read and agreed to the published version of the manuscript.

Funding: This research received no external funding.

Informed Consent Statement: Informed consents were signed antemortem by those participating in the study, stating that they knowingly and willingly donate their bodies to medical education and scientific work.

Data Availability Statement: Authors declare that all related data are available concerning researchers by the corresponding author's email.

Acknowledgments: The authors sincerely thank those who donated their bodies to science so that anatomical research could be performed. Results from such research can potentially increase mankind's overall knowledge that can then improve patient care. Therefore, these donors and their families deserve our highest gratitude [25]. The authors wish to thank Niko Valnarov, Marina Klissourova, Assia Konsulova, and Georgi Kostov for their technical support. The authors wish to thank Miglena Nevyanova Drincheva for her beautiful artwork.

Conflicts of Interest: The authors declare no conflict of interest.

References

1. Singh, K.; Fares, R. Pelvic and Para-aortic Lymphadenectomy in Gynecologic Cancers. In *Gynecologic and Obstetric Surgery: Challenges and Management Options*, 1st ed.; Coomarasamy, A., Shafi, M., Davila, G.W., Chan, K.K., Eds.; John Wiley & Sons Inc.: New York, NY, USA, 2016; pp. 408–411. [CrossRef]
2. Mangan, C.E.; Rubin, S.C.; Rabin, D.S.; Mikuta, J.J. Lymph node nomenclature in gynecologic oncology. *Gynecol. Oncol.* **1986**, *23*, 222–226. [CrossRef]
3. Bakkum Gamez, J.N. Lymphadenectomy in the Management of Gynecologic Cancer. *Clin. Obstet. Gynecol.* **2019**, *62*, 749–755. [CrossRef]
4. Fowler, J.; Backes, F. Pelvic and Paraaortic Lymphadenectomy in Gynecologic Cancers up to Date 2020. Available online: https://www.uptodate.com/contents/pelvic-and-paraaortic-lymphadenectomy-in-gynecologiuc-cancers (accessed on 8 November 2020).
5. Cibula, D.; Abu-Rustum, N.R. Pelvic lymphadenectomy in cervical cancer—surgical anatomy and proposal for a new classification system. *Gynecol. Oncol.* **2010**, *116*, 33–37. [CrossRef]
6. Canessa, C.E.; Miegge, L.M.; Bado, J.; Silveri, C.; Labandera, D. Anatomic Study of Lateral Pelvic Lymph Nodes: Implications in the Treatment of Rectal Cancer. *Dis. Colon Rectum* **2004**, *47*, 297–303. [CrossRef]
7. Selçuk, İ.; Uzuner, B.; Boduç, E.; Baykuş, Y.; Akar, B.; Güngör, T. Pelvic lymphadenectomy: Step-by-step surgical education video. *J. Turk. Ger. Gynecol. Assoc.* **2020**, *21*, 66–69. [CrossRef]
8. McMahon, C.J.; Rofsky, N.M.; Pedrosa, I. Lymphatic Metastases from Pelvic Tumors: Anatomic Classification, Characterization, and Staging. *Radiology* **2010**, *254*, 31–46. [CrossRef]
9. Kolbenstvedt, A.; Kolstad, P. The difficulties of complete pelvic lymph node dissection in radical hysterectomy for carcinoma of the cervix. *Gynecol. Oncol.* **1976**, *4*, 244–254. [CrossRef]
10. Ovcharov, V.; Vankov, V. *Human Anatomy*, 14th ed.; ARSO Publishing: Sofia, Bulgaria, 2019; pp. 632–633.
11. Vincent, P. *Lymphatic System 9 Quick Study Academic*, 1st ed.; BarCharts Publishing: Boca Raton, FL, USA, 2016.
12. Paño, B.; Sebastià, C.; Ripoll, E.; Paredes, P.; Salvador, R.; Buñesch, L.; Nicolau, C. Pathways of Lymphatic Spread in Gynecologic Malignancies. *RadioGraphics* **2015**, *35*, 916–945. [CrossRef] [PubMed]
13. Sakuragi, N.; Satoh, C.; Takeda, N.; Hareyama, H.; Takeda, M.; Yamamoto, R.; Fujimoto, T.; Oikawa, M.; Fujino, T.; Fujimoto, S. Incidence and distribution pattern of pelvic and paraaortic lymph node metastasis in patients with stages IB, IIA, and IIB cervical carcinoma treated with radical hysterectomy. *Cancer* **1999**, *85*, 1547–1554. [CrossRef]
14. Liu, Z.; Hu, K.; Liu, A.; Shen, J.; Hou, X.; Lian, X.; Sun, S.; Yan, J.; Zhang, F. Patterns of lymph node metastasis in locally advanced cervical cancer. *Medicine* **2016**, *95*, e4814. [CrossRef] [PubMed]
15. Panici, P.B.; Scambia, G.I.; Baiocchi, G.A.; Matonti, G.I.; Capelli, A.R.; Mancuso, S.A. Anatomical study of para-aortic and pelvic lymph nodes in gynecologic malignancies. *Obstet. Gynecol.* **1992**, *79*, 498–502. [PubMed]
16. Centini, G.; Fernandes, R.P.; Afors, K.; Murtada, R.; Puga, M.F.; Wattiez, A. Radical Hysterectomy and Pelvic Lymphadenectomy (French School). *Hysterectomy* **2017**, 589–595. [CrossRef]
17. Benedetti Panici, P.; Basile, S.; Angioli, R. Pelvic and aortic lymphadenectomy in cervical cancer: The standardization of surgical procedure and its clinical impact. *Gynecol. Oncol.* **2009**, *113*, 284–290. [CrossRef]
18. Jones, H.W.; Rock, J.A. *Te Linde's Operative Gynecology*, 10th ed.; Lippincott Williams & Wilkins: Philadelphia, PA, USA, 2008.
19. Chen, J.J.; Zhu, Z.S.; Zhu, Y.Y.; Shi, H.Q. Applied anatomy of pelvic lymph nodes and its clinical significance for prostate cancer:a single-center cadaveric study. *BMC Cancer* **2020**, *20*, 330. [CrossRef]
20. Frober, R. Surgical anatomy of the ureter. *BJU Int.* **2007**, *100*, 949–965. [CrossRef]
21. Razvan, R.; Geiorgescu, D.; Geavlete, P.; Bogdan, B. Retrograde Ureteroscopy: Handbook of Endourology. In *Notions of Histology, Anatomy, and Physiology of the Upper Urinary Tract*, 1st ed.; Geavlete, P., Ed.; Academic Press: Cambridge, MA, USA, 2016; pp. 7–19.
22. Wedel, T. Topographical Anatomy for Hysterectomy Procedures. In *Hysterectomy a Comprehensive Surgical Approach*, 1st ed.; Alkatout, I., Mettler, L., Eds.; Springer: Cham, Switzerland, 2018; pp. 37–60.
23. Selçuk, İ.; Ersak, B.; Tatar, İ.; Güngör, T.; Huri, E. Basic clinical retroperitoneal anatomy for pelvic surgeons. *Turk. J. Obstet. Gynecol.* **2018**, *15*, 259–269. [CrossRef]
24. Amar, A.D.; Hutch, J.A. Anomalies of the Ureter. In *Malformations (Handbuch der Urologie/Encyclopedia of Urology/Encyclopédie d'Urologie)*; Springer: Berlin, Germany, 1968. [CrossRef]
25. Nation, E.F. Duplication of the Kidney and Ureter: A Statistical Study of 230 New Cases. *J. Urol.* **1944**, *51*, 456–465. [CrossRef]
26. Foley, C.E.; Mansuria, S. Ureteral anomalies in gynecologic surgery. *J. Minim. Invasive Gynecol.* **2020**, *27*, 566–567. [CrossRef] [PubMed]
27. Mayers, M.M. Diverticulum of the ureter. *J. Urol.* **1949**, *61*, 344–350. [CrossRef]
28. McLoughlin, L.C.; Davis, N.F.; Dowling, C.; Eng, M.P.; Power, R.E. Ureteral diverticulum: A review of the current literature. *Can. J. Urol.* **2013**, *20*, 6893–6896.
29. Papin, E.; Eisendrath, D.N. Classification of Renal and Ureteral Anomalies. *Ann. Surg.* **1927**, *85*, 735–756. [PubMed]
30. Tawfik, A.M.; Younis, M.H. Computed Tomography Imaging Appearance of a Unique Variant of Retroiliac Ureter. *Urology* **2016**, *88*, e7–e9. [CrossRef]
31. Prasad, H.L.; Karthikeyan, V.S.; Shivalingaiah, M.; Ratkal, C.S. Retroiliac ureter presenting as right upper ureteric obstruction—Report of a rare case. *Int. J. Res. Med. Sci.* **2015**, *3*, 2143–2144. [CrossRef]

32. Ebiloglu, T.; Kaya, E.; Zorba, O.; Kibar, Y.; Gok, F. Retro-Iliac Ureters: Review of Literature and Two Cases with Two Different Techniques. *Turk. Klin. J. Case Rep.* **2016**, *24*, 66–69. [CrossRef]
33. Parashar, M.; Sharma, R.K.; Kumar, S. Ureteric injury in gynaecological surgery: A rare but serious event. *Int. J. Health Sci. Res.* **2019**, *9*, 397–401.
34. Mylonas, I.; Briese, V.; Vogt-Weber, B.; Friese, K. Complete bilateral crossed ureteral duplication observed during a radical hysterectomy with pelvic lymphadenectomy for ovarian cancer. A case report. *Arch. Gynecol. Obstet.* **2003**, *267*, 250–251. [CrossRef]
35. Ostrzenski, A.; Radolinski, B.; Ostrzenska, K.M. A review of laparoscopic ureteral injury in pelvic surgery. *Obstet. Gynecol. Surv.* **2003**, *58*, 794–799. [CrossRef]
36. Davis, A.A. Transection of Duplex Ureter During Vaginal Hysterectomy. *Cureus* **2020**, *12*, e6597. [CrossRef]
37. Benedetti-Panici, P.; Maneschi, F.; Scambia, G.; Greggi, S.; Mancuso, S. Anatomic abnormalities of the retroperitoneum encountered during aortic and pelvic lymphadenectomy. *Am. J. Obstet. Gynecol.* **1994**, *170 Pt 1*, 111–116. [CrossRef]
38. Dalzell, A.P.B.; Robinson, R.G.; Crooke, K.M. Duplex ureter damaged during laparoscopic hysterectomy. *Pelviperineology* **2010**, *29*, 88.
39. Subbiah, S.; Rajaraman, N.N. Congenital anomalies in surgical oncology practice. *Edorium J. Surg.* **2015**, *2*, 29–36. [CrossRef]
40. Li, J.; Wang, Z.; Chen, C.; Liu, P.; Duan, H.; Chen, L.; Wang, J.; Tan, H.; Li, P.; Zhao, C.; et al. Distribution of iliac veins posterior to the common iliac artery bifurcation related to pelvic lymphadenectomy: A digital in vivo anatomical study of 442 Chinese females. *Gynecol. Oncol.* **2016**, *141*, 538–542. [CrossRef]
41. Gray, H.; Standring, S.; Hrold Ellis, H.; Berkovitz, B. *Gray's Anatomy: The Anatomical Basis of Clinical Practice*, 39th ed.; Elsevier Churchill Livingstone Edinburgh: New York, NY, USA, 2005; pp. 2171–2175.
42. Tubbs, R.S.; Shoja, M.M.; Loukas, M. *Bergman's Comprehensive Encyclopedia of Human Anatomic Variation*; John Wiley & Sons: Hoboken, NJ, USA, 2016; pp. 669–673, 884–889.
43. Shetty, S.; Kantha, L.; Sheshgiri, C. Bilateral absence of common iliac artery—A cadaveric observation. *Int. J. Anat. Var.* **2013**, *6*, 7–8.
44. Dabydeen, D.A.; Shabashov, A.; Shaffer, K. Congenital Absence of the Right Common Iliac Artery. *Radiol. Case Rep.* **2015**, *3*, 47. [CrossRef]
45. Llauger, J.; Sabaté, J.M.; Guardia, E.; Escudero, J. Congenital absence of the right common iliac artery: CT and angiographic demonstration. *Eur. J. Radiol.* **1995**, *21*, 128–130. [CrossRef]
46. Rusu, M.C.; Cergan, R.; Dermengiu, D.; Curcă, G.C.; Folescu, R.; Motoc, A.G.M.; Jianu, A.M. The iliolumbar artery—Anatomic considerations and details on the common iliac artery trifurcation. *Clin. Anat.* **2009**, *23*, 93–100. [CrossRef]
47. Lierse, W. *Applied Anatomy of the Pelvis*, 1st ed.; Springer: Berlin/Heidelberg, Germany, 1987; pp. 47–58. [CrossRef]
48. Badagabettu Nayak, S.; Padur Aithal, A.; Kumar, N.; Regunathan, D.; Shetty, P.; Alathady Maloor, P. A cadaveric study of variations of external iliac artery and its implication in trauma and radiology. *Morphologie* **2019**, *103*, 24–31. [CrossRef]
49. Tamisier, D.; Melki, J.-P.; Cormier, J.-M. Congenital Anomalies of the External Iliac Artery: Case Report and Review of the Literature. *Ann. Vasc. Surg.* **1990**, *4*, 510–514. [CrossRef]
50. Kawashima, T.; Sato, K.; Sasaki, H. A human case of hypoplastic external iliac artery and its collateral pathways. *Folia Morphol.* **2006**, *65*, 157–160.
51. Okamoto, K.; Wakebe, T.; Saiki, K.; Nagashima, S. Consideration of the potential courses of the common iliac artery. *Anat. Sci. Int.* **2005**, *80*, 116–119. [CrossRef]
52. Safi, K.C.; Teber, D.; Moazen, M.; Anghel, G.; Maldonado, R.V.; Rassweiler, J.J. Laparoscopic repair of external iliac-artery transection during laparoscopic radical prostatectomy. *J. Endourol.* **2006**, *20*, 237–239. [CrossRef] [PubMed]
53. Boonruangsri, P.; Suwannapong, I.; Rattanasuwan, S.; Iamsaard, S. Aneurysm, tortuosity, and kinking of abdominal aorta and iliac arteries in Thai cadavers. *Int. J. Morphol.* **2015**, *33*, 73–76. [CrossRef]
54. Moul, J.W.; Wind, G.G.; Wright, C.R. Tortuous and aberrant external iliac artery precluding radical retropubic prostatectomy for prostate cancer. *Urology* **1993**, *42*, 450–452. [CrossRef]
55. Kostov, S.; Slavchev, S.; Dzhenkov, D.; Stoyanov, G.; Dimitrov, N.; Yordanov, A. Corona mortis, aberrant obturator vessels, accessory obturator vessels: Clinical applications in gynecology. *Folia Morphol.* **2020**. [CrossRef]
56. Sañudo, J.R.; Roig, M.; Rodriguez, A.; Ferreira, B.; Domenech, J.M. Rare origin of the obturator, inferior epigastric and medial circumflex femoral arteries from a common trunk. *J. Anat.* **1993**, *183*, 161–163.
57. Fătu, C.; Puişoru, M.; Fătu, I.C. Morphometry of the internal iliac artery in different ethnic groups. *Ann. Anat.* **2006**, *188*, 541–546. [CrossRef]
58. Adachi, B. *Das Arteriensystem der Japaner, Bd. II*; Kyoto. Supp. to Acta Scholae Medicinalis Universitatis Imperalis in Kioto: Tokyo, Japan, 1928; Volume 9, pp. 1926–1927.
59. Yamaki, K.; Saga, T.; Doi, Y.; Aida, K.; Yoshizuka, M. A statistical study of the branching of the human internal iliac artery. *Kurume Med. J.* **1998**, *45*, 333–340. [CrossRef]
60. Liapis, K.; Tasis, N.; Tsouknidas, I.; Tsakotos, G.; Skandalakis, P.; Vlasis, K.; Filippou, D. Anatomic variations of the Uterine Artery. Review of the literature and their clinical significance. *Turk. J. Obstet. Gynecol.* **2020**, *17*, 58–62. [CrossRef]
61. Kumari, S.; Trinesh Gowda, M.S. A study of variations of origin of obturator artery: Review in south Indian population. *J. Anat. Soc. India* **2016**, *65*, S1–S4. [CrossRef]

62. Granite, G.; Meshida, K.; Wind, G. Frequency and Clinical Review of the Aberrant Obturator Artery: A Cadaveric Study. *Diagnostics* **2020**, *10*, 546. [CrossRef]
63. Sañudo, J.R.; Mirapeix, R.; Rodriguez-Niedenführ, M.; Maranillo, E.; Parkin, I.G.; Vázquez, T. Obturator artery revisited. *Int. Urogynecol. J.* **2011**, *22*, 1313–1318. [CrossRef]
64. Bae, J.W.; Lee, J.H.; Choi, J.S.; Son, C.E.; Jeon, S.W.; Hong, J.H.; Eom, J.M.; Joo, K.J. Laparoscopic lymphadenectomy for gynecologic malignancies: Evaluation of the surgical approach and outcomes over a seven-year experience. *Arch Gynecol. Obstet.* **2012**, *285*, 823–829. [CrossRef]
65. Ricciardi, E.; di Martino, G.; Maniglio, P.; Schimberni, M.; Frega, A.; Jakimovska, M.; Kobal, B.; Moscarini, M. Life-threatening bleeding after pelvic lymphadenectomy for cervical cancer: Endovascular management of ruptured false aneurysm of the external iliac artery. *World J. Surg. Onc.* **2012**, *10*, 149. [CrossRef]
66. Ishikawa, M.; Nakayama, K.; Razia, S.; Yamashita, H.; Ishibashi, T.; Sato, S.; Sasamori, H.; Sawada, K.; Kurose, S.; Ishikawa, N.; et al. External Iliac Artery Injury and Thrombosis during Laparoscopic Gynecologic Surgery. *Case Rep. Obstet. Gynecol.* **2020**. [CrossRef]
67. Gyimadu, A.; Salman, M.C.; Karcaaltincaba, M.; Yuce, K. Retroperitoneal vascular aberrations increase the risk of vascular injury during lymphadenectomy in gynecologic cancers. *Arch Gynecol. Obstet.* **2012**, *286*, 449–455. [CrossRef] [PubMed]
68. Fotopoulou, C.; Neumann, U.; Kraetschell, R.; Lichtenegger, W.; Sehouli, J. External iliac artery ligation due to late postoperative rupture after radical lymphadenectomy for advanced ovarian cancer—Two case reports. *Eur. J. Gynaecol. Oncol.* **2010**, *31*, 198–200. [PubMed]
69. Mehta, K.; Iwanaga, J.; Tubbs, R.S. Absence of the Right Common Iliac Vein with the Right Internal Iliac Vein Arising from the Left Common Iliac Vein: Case Report. *Cureus* **2019**, *11*, e4575. [CrossRef]
70. Morris, H. *Human Anatomy: A Complete Systematic Treatise by English and American Authors*, 5th ed.; P. Blakiston, Son & Co.: Philadelphia, PA, USA, 1893. [CrossRef]
71. Cardinot, T.M.; Aragão, A.H.; Babinski, M.A.; Favorito, L.A. Rare variation in course and affluence of internal iliac vein due to its anatomical and surgical significance. *Surg. Radiol. Anat.* **2006**, *28*, 422–425. [CrossRef]
72. Yahyayev, A.; Bulakci, M.; Yilmaz, E.; Ucar, A.; Sayin, O.A.; Yekeler, E. Absence of the right iliac vein and an unusual connection between both common femoral veins. *Phlebology* **2013**, *28*, 162–164. [CrossRef]
73. Panchal, P.; Chaturvedi, H. Agenesis of common iliac vein encroaching development of inferior vena cava. *IJAV* **2014**, *7*, 21–23.
74. Nusrath, S.; Pawar, S.; Goel, V.; Raju, K. Common Internal Iliac Vein Joining Inferior Vena Cava—A Rare Anatomical Variation: Anatomic and Surgical Relevance. *Indian J. Surg.* **2020**, *82*, 701–703. [CrossRef]
75. Kose, M.F.; Turan, T.; Karasu, Y.; Gundogdu, B.; Boran, N.; Tulunay, G. Anomalies of Major Retroperitoneal Vascular Structure. *Int. J. Gynecol. Cancer* **2011**, *21*, 1312–1319. [CrossRef] [PubMed]
76. Jasani, V.; Jaffray, D. The anatomy of the iliolumbar vein: A cadaver study. *J. Bone Joint Surg. Br.* **2002**, *84*, 1046–1049. [CrossRef]
77. Lolis, E.; Panagouli, E.; Venieratos, D. Study of the ascending lumbar and iliolumbar veins: Surgical anatomy, clinical implications and review of the literature. *Ann. Anat.* **2011**, *193*, 516–529. [CrossRef]
78. Unruh, K.P.; Camp, C.L.; Zietlow, S.P.; Huddleston, P.M. Anatomical variations of the iliolumbar vein with application to the anterior retroperitoneal approach to the lumbar spine: A cadaver study. *Clin. Anat.* **2008**, *21*, 666–673. [CrossRef]
79. Venieratos, D.; Panagouli, E.; Lolis, E. Variations of the iliac and pelvic venous systems with special attention to the drainage patterns of the ascending lumbar and iliolumbar veins. *Ann. Anat.* **2012**, *194*, 396–403. [CrossRef]
80. Hamid, M.; Toussaint, P.J.; Delmas, V.; Gillot, C.; Coutaux, A.; Plaisant, O. Anatomical and radiological evidence for the iliolumbar vein as an inferior lumbar venous system. *Clin. Anat.* **2007**, *20*, 545–552. [CrossRef]
81. Sivakumar, G.; Paluzzi, A.; Freeman, B. Avulsion of ascending lumbar and iliolumbar veins in anterior spinal surgery: An anatomical study. *Clin. Anat.* **2007**, *20*, 553–555. [CrossRef]
82. Lotz, P.R.; Seeger, J. Normal variation in iliac venous anatomy. *AJR* **1982**, *138*, 735–738. [CrossRef] [PubMed]
83. Djedovic, G.; Putz, D. Case report: Description of a venous annulus of the external iliac vein. *Ann. Anat.* **2006**, *188*, 451–453. [CrossRef] [PubMed]
84. Hayashi, S.; Naito, M.; Yakura, T.; Kumazaki, T.; Itoh, M.; Nakano, T. A case of an additional right external iliac vein surrounding the right external iliac artery and lacking the right common iliac vein. *Anat. Sci. Int.* **2015**, *91*, 106–109. [CrossRef] [PubMed]
85. Nezhat, C.; Childers, J.; Nezhat, F.; Nezhat, C.H.; Seidman, D.S. Major retroperitoneal vascular injury during laparoscopic surgery. *Hum. Reprod.* **1997**, *12*, 480–483. [CrossRef] [PubMed]
86. Herraiz Roda, J.L.; Llueca, A.; Maazouzi, Y.; Piquer Simó, D.; Guijarro Colomer, M.; Sentís Masllorens, J. Complications of laparoscopic lymphadenectomy for gynecologic malignancies. Experience of 372 patients. *Res. Rep. Gynaecol. Obstet.* **2017**, *1*, 12–16.
87. Ghassemi, A.; Furkert, R.; Prescher, A.; Riediger, D.; Knobe, M.; O'dey, D.; Gerressen, M. Variants of the supplying vessels of the vascularized iliac bone graft and their relationship to important surgical landmarks. *Clin. Anat.* **2013**, *26*, 509–521. [CrossRef]
88. Elsy, B.; Alghamdi, A.; Osman, L. Bilateral branching variants of internal and external iliac arteries—Cadaveric study Case Report. *Eur. J. Anat.* **2020**, *24*, 63–68.
89. Vidal, V.; Monnet, O.; Jacquier, A.; Bartoli, J.-M.; Tropiano, P. Accessory Iliac Vein. *J. Spinal Disord Tech.* **2010**, *23*, 398–403. [CrossRef]

90. Shin, M.; Lee, J.B.; Park, S.B.; Park, H.J.; Kim, Y.S. Multidetector computed tomography of iliac vein variation: Prevalence and classification. *Surg. Radiol. Anat.* **2014**, *37*, 303–309. [CrossRef]
91. Morita, S.; Saito, N.; Mitsuhashi, N. Variations in internal iliac veins detected using multidetector computed tomography. *Acta Radiol.* **2007**, *48*, 1082–1085. [CrossRef]
92. Chong, G.O.; Lee, Y.H.; Hong, D.G.; Cho, Y.L.; Lee, Y.S. Anatomical variations of the internal iliac veins in the presacral area: Clinical implications during sacral colpopepxy or extended pelvic lymphadenectomy. *Clin. Anat.* **2014**, *28*, 661–664. [CrossRef]
93. Nayak, S.B. Dangerous twisted communications between external and internal iliac veins which might rupture during catheterization. *Anat. Cell. Biol.* **2018**, *51*, 309–311. [CrossRef]
94. Kanjanasilp, P.; Ng, J.L.; Kajohnwongsatit, K.; Thiptanakit, C.; Limvorapitak, T.; Sahakitrungruang, C. Anatomical Variations of Iliac Vein Tributaries and Their Clinical Implications During Complex Pelvic Surgeries. *Dis. Colon Rectum.* **2019**, *62*, 809–814. [CrossRef] [PubMed]
95. Duan, H.; Liu, P.; Chen, C.; Chen, L.; Li, P.; Li, W.; Gong, S.; Xv, Y.; Chen, R.; Tang, L. Reconstruction of three-dimensional vascular models for lymphadenectomy before surgery. *Minim. Invasive Ther. Allied Technol.* **2020**, *29*, 42–48. [CrossRef] [PubMed]
96. Sanna, B.; Henry, B.M.; Vikse, J.; Skinningsrud, B.; Pękala, J.R.; Walocha, J.A.; Cirocchi, R.; Tomaszewski, K.A. The prevalence and morphology of the corona mortis (Crown of death): A meta-analysis with implications in abdominal wall and pelvic surgery. *Injury* **2018**, *49*, 302–308. [CrossRef] [PubMed]
97. Darmanis, S.; Lewis, A.; Mansoor, A.; Bircher, M. Corona mortis: An anatomical study with clinical implications in approaches to the pelvis and acetabulum. *Clin. Anat.* **2007**, *20*, 433–439. [CrossRef] [PubMed]
98. Berberoĝlu, M.; Uz, A.; Özmen, M.M.; Bozkurt, C.; Erkuran, C.; Taner, S.; Tekin, A.; Tekdemir, I. Corona mortis: An anatomic study in seven cadavers and an endoscopic study in 28 patients. *Surg. Endosc.* **2001**, *15*, 72–75. [CrossRef] [PubMed]
99. Lee, Y.S. Early experience with laparoscopic pelvic lymphadenectomy in women with gynecologic malignancy. *J. Am. Assoc. Gynecol. Laparosc.* **1999**, *6*, 59–63. [CrossRef]
100. Selçuk, İ.; Tatar, İ.; Fırat, A.; Huri, E.; Güngör, T. Is corona mortis a historical myth? A perspective from a gynecologic oncologist. *J. Turk. Ger. Gynecol. Assoc.* **2018**, *19*, 171–172. [CrossRef]
101. Anagnostopoulou, S.; Kostopanagiotou, G.; Paraskeuopoulos, T.; Chantzi, C.; Lolis, E.; Saranteas, T. Anatomic Variations of the Obturator Nerve in the Inguinal Region. *Reg. Anesth. Pain Med.* **2009**, *34*, 33–39. [CrossRef]
102. Anloague, P.A.; Huijbregts, P. Anatomical Variations of the Lumbar Plexus: A Descriptive Anatomy Study with Proposed Clinical Implications. *J. Man Manip. Ther.* **2009**, *17*, 107E–114E. [CrossRef]
103. Turgut, M.; Protas, M.; Gardner, B.; Oskouian, R.; Loukas, M.; Tubbs, R. The accessory obturator nerve: An anatomical study with literature analysis. *Anatomy* **2017**, *11*, 121–127. [CrossRef]
104. Archana, B.J.; Nagaraj, D.N.; Pradeep, P.; Subhash, L. Anatomical Variations of Accessory Obturator Nerve: Cadaveric Study with Proposed Clinical Implications. *Int. J. Anat. Res.* **2016**, *4*, 2158–2161. [CrossRef]
105. Gindha, G.S.; Arora, D.; Kaushal, S.; Chhabra, U. Variations in origin of the genitofemoral nerve from the lumbar plexuses in north Indian population (a cadaveric study). *MOJ Anat. Physiol.* **2015**, *1*, 72–76. [CrossRef]
106. Paul, L.; Shastri, D. Anatomical variations in formation and branching pattern of the border nerves of lumbar Region. *Natl. J. Clin. Anat.* **2019**, *8*, 57–61. [CrossRef]
107. Deepti, A.; Shyam, S.T.; Subhash, K.; Usha, C. Morphology of Lumbar Plexus and Its Clinicalsignificance. *Int. J. Anat. Res.* **2016**, *4*, 2007–2014. [CrossRef]
108. Cardosi, R.; Cox, C.; Hoffman, M. Postoperative neuropathies after major pelvic surgery. *Obstet. Gynecol.* **2002**, *100*, 240–244. [CrossRef] [PubMed]

Article

Anatomographic Variants of Sphenoid Sinus in Ethiopian Population

Tizita K. Degaga [1], Abay M. Zenebe [2], Amenu T. Wirtu [2], Tequam D. Woldehawariat [3], Seife T. Dellie [3] and Jicksa M. Gemechu [4,*]

1 Department of Biomedical Sciences, Menelik II Health Science College, Kotebe Metropolitan University, Addis Ababa 31228, Ethiopia; kinfetizita@gmail.com
2 Department of Anatomy, College of Health Sciences, Addis Ababa University, Addis Ababa 9086, Ethiopia; abaymulu@gmail.com (A.M.Z.); amannut2002@gmail.com (A.T.W.)
3 Department of Radiology, College of Health Sciences, Addis Ababa University, Addis Ababa 9086, Ethiopia; tequamrad2018@gmail.com (T.D.W.); seifeteferi@gmail.com (S.T.D.)
4 Department of Foundational Medical Studies, Oakland University William Beaumont School of Medicine, Detroit, MI 48309, USA
* Correspondence: gemechu@oakland.edu; Tel.: +1-(248)-370-3667; Fax: +1-(248)-370-4060

Received: 26 August 2020; Accepted: 14 October 2020; Published: 19 November 2020

Abstract: Neurosurgeons often neglect the sphenoid sinus due to its deep location and difficulties in accessing during surgical interventions. Disease of the sphenoid sinus is difficult to diagnose since its presenting symptoms are difficult to recognize. Moreover, compared with other paranasal sinuses, the sphenoid sinus is considered the most variable air sinus in terms of its degree of pneumatization, number and position of inter-sinus septa, and its relationship with the surrounding anatomical structures. Anatomical variations of the sphenoid sinus are significant from a neurosurgical point of view. Understanding of these variations and its relationships with surrounding structures such as the internal carotid artery, optic nerve, and pituitary gland are clinically relevant to minimize injuries associated with surgical procedures that involve sphenoid sinus. We implemented principles of imaging using computed tomography to elucidate any anatomical variations of the sphenoid sinus in the Ethiopian population. We conducted a prospective study in 200 patients with ages 18–79, who underwent scans of the sphenoid sinus at the Tikur Anbessa Referral Teaching Hospital in 2017–2018. Our findings revealed an incidence of anatomographical variations in terms of pneumatization that varied between 2–50%. These variants include 2% conchal, 25.5% presellar, 50% sellar, and 22.5% postsellar pneumatization. We also demonstrated anatomographic variants in terms of septation, 77.5% single complete septa, 11.5% single incomplete, 10% double septa, and 1% absence of septa. In summary, the sellar pneumatization was found to be the most clinically relevant anatomographic variant among Ethiopians participating in the study, of which 90% were tomographically single septated. These variants must be taken into consideration during trans-sphenoidal surgery and knowledge of the variations has clinical implication in minimizing injuries during invasive surgical procedures involving the sphenoid sinus.

Keywords: anatomographical variation; sphenoid sinus; septation; pneumatization; Ethiopian population

---

1. Introduction

The sphenoid sinus is the most inaccessible paranasal sinus, enclosed within the body of the sphenoid bone and intimately related to numerous neurovascular and glandular structures [1]. The failure of surgeons to understand the racial variations of the anatomical landmarks of the sphenoid sinus is often described as a potential risk factor in clinical interventions [2]. The presence of Onodi cells may be accompanied by morphological variations of the neighboring anatomical structures [3]. Such anatomographic variants carry significant surgical implications in sinonasal regions.

Computerized tomography (CT) is an imaging modality used for diagnosing diseases and evaluating injuries. It also plays an important role in the diagnosis of anatomical variations, which has relevant implications in clinical decision making during surgical interventions. CT of the paranasal sinuses reveals a wide spectrum of findings associated with the normal pneumatization processes inside the sinus cavities and in the adjacent marrow spaces [4]. The sphenoid sinus received clinical significance after neurosurgeons discovered the trans-sphenoid approach for pituitary tumor surgeries in the new era of minimal invasive surgery [5]. These mucous membrane lined air cells are situated within the body of the sphenoid bone, communicating with the roof of the nasal cavity through an opening into the sphenoethmoidal recess. It is closely related with the surrounding vital structures such as optic chiasm, cavernous sinus, pituitary gland, and internal carotid arteries [2,6].

When compared to other paranasal sinuses, the sphenoid sinus follows a different developmental pattern [7]. Its formation takes place in the body of sphenoid bone, beginning as an invagination of the nasal mucosa into the posterior portion of the cartilaginous nasal capsule between the third and fourth months of fetal life. A recess appears at birth, which is present between the presphenoid body and the sphenoid concha. After birth, the sphenoid sinus exists primarily as a pit in the sphenoethmoid recess; by age 3, starts pneumatization of the sphenoid bone; by age 7, extended toward the sella turcica; and it reaches the final form in the mid-teens. Although the definitive cavity forms at puberty, the actual sinus cavity starts becoming visible between 8–10 years. On the posterior nasal wall, the origin of the sphenoid sinus can be clearly identified by the location of its ostium [8]. The paired sinuses generally develop asymmetrically, separated by the inter-sinus bony septum [7,9]. During childhood, maturation of the bone from red to yellow marrow takes place in the anterior part of the sphenoid bone [10].

The sphenoid sinus is considered the most variable of the paranasal sinuses in terms of degree and type of pneumatization, number and position of inter-sinus septa, relationship with the surrounding structures like the cranial nerves (CN II, III, IV, V, VI), the internal carotid artery inside the cavernous sinus, and the pituitary gland [11,12]. The relationship of the sphenoid sinus is a prerequisite to safe and effective surgical treatment for lesions in the nasal region, and lack of orientation during dissection might lead to different surgical complications [8]. The sphenoid sinus drains directly into the nasal cavity by the sphenoethmoid recess [13]. Knowledge of anatomographic variations of the sphenoid sinus and its anatomic relationship with surrounding structures shorten the duration of surgeries while reducing further comorbid complications. Radiological evaluation of anatomographical variations of sphenoid sinus assist to understand the key steps taken during tumor progression as well as guide surgeons in the management of the relevant pathologies [14].

2. Materials and Methods

A prospective observational study was conducted at the Tikur Anbessa Specialized Teaching Hospital, Addis Ababa University, on 200 patients (117 female and 83 male) who underwent CT scan imaging for paranasal sinuses evaluation during 2017–2018. Ethical clearance was obtained from both, Department of Anatomy and Radiology Research Ethics Review Committee, DRERC/01/09, 06/01/2017. CT scan images were acquired from patients with informed consent. Data were anonymous to maintain patient confidentiality. Images with a slice thickness > 3 mm, low resolution quality, and those with metallic artifacts that impair sinus visualization were excluded from the study. In addition, patients with a history of prior sinus or sphenoid surgery, facial trauma, and obscured sphenoid sinus pathology were also excluded. A Philips 128 slice MDCT scanner, 130 kV, 120 mAs, was used to acquire images. Images were taken in the axial planes and then 1.25 mm slices were reconfigured into coronal and sagittal planes [15]. Patients were placed in the supine position with their chin hyperextended and scan plane angled perpendicular to the hard palate. Axial scans were performed from the maxillary sinus floor to the level of the frontal sinus roof, in a plane parallel with the hard palate [10]. Images were reviewed on the console with varying window levels and widths. The sphenoid sinuses were reviewed in both axial and coronal planes, and the total number of septa were counted (single, double, and absent) and compared in both planes [16]. The data was processed by Maxi and RadiAnt viewer computer softwares (Mission Viejo, CA 92691, USA and 5.5.1.23267, Medixant, Pozan', Poland respectively). Selected images for the study were reviewed and interpreted by one neuroradiologist and five senior radiology residents. For detailed anatomographic variations, only findings consistently reported by all were used. The quantitative data was captured by an Excel spreadsheet, cleaned, and exported to SPSS version 20 for analysis. The mean, median, frequency distribution and proportion of variables indicating variation were reported using descriptive statistics. A p-value < 0.05 was considered statically significant based on the Chi-square tests.

3. Results

In this study, 200 patients participated, out of which 117 (58.5%) were females, 83 (41.5%) were males, and seven (3.4%) were non-respondents. The participants' age ranged from 18 to 79, with a mean of (±SD) 43 (±14.5) years. The mean age of males was 51, while for females was 37 (Table 1). Our findings showed that anatomographic variants of sphenoid sinus were 100 (50%) seller, 51 (25.5%) presellar, 45 (22.5%) postsellar, and 4 (2%) were conchal types of pneumatization (Figures 1 and 2; Table 2). We also found that pneumatization of the anterior clinoid process (ACP) was 36 (18%), pterygoid plates (PP) was 3 (15%), and the greater wing of sphenoid was (GWS) 33 (16.5%) (Figure 3).

Septal bone septation of the sphenoid sinus in the Ethiopian population was single 179 (89%). That is, 115 (77.5%) complete and 23 (11.5%) incomplete), 20 (10%) double (all are complete), and two (1%) were either null or without septation (Figures 4 and 5; Table 2).

We also investigated the dehiscence and protrusion of sphenoid sinus in relation to internal carotid artery, optic nerve, and foramen rotundum (V2) (Table 3). Protrusion of the internal carotid artery into the sphenoid sinus was identified on the CT images of 37 (18.5%) patients: the right side alone was involved in 12 (32.4%) patients; the left alone in nine (24.3%) patients, and bilateral involvement was involved in 16 (43.2%) patients. The dehiscence of the bony sphenoidal wall of the internal carotid artery occurred in 24 (12%) patients; the right side alone was involved in seven (29.17%) patients, left side alone in 10 (41.7%) patients, and bilateral involvement was observed in seven (29.17%) patients. Nineteen (9.5%) cases had the optic nerve protrusion into the sphenoid sinus: right sided in one (5.26%) case, left side in 10 (52.63%) cases, and bilateral involvement in eight (42.1%) cases. However, dehiscence occurred in 31 (15.5%) patients; right sided in 12 (38.7%) cases, left side in 9 (29%) cases, and bilateral involvement in 10 (32.5%) cases. From the total participants, 25 (12.5%) cases had maxillary nerve protrusion into the sphenoid sinus; the right side alone was involved in seven (28%) patients; the left alone in eight (32%) patients, and bilateral involvement was involved in 10 (40%) patients. However, dehiscence occurred in 25 (12.5%) patients; right sided in four (16%) cases, left side in 11 (44%) cases, and bilateral involvement in 10 (40%) cases. We also identified the presence of onodi cells only in two (1%) cases (Figure 6). The Pearson Chi-square-χ^2 test for the pterygoid plate and vidian canal dehiscence, dehiscence of the optic nerve, and pneumatization of the anterior clinoid process at the 95% confidence interval was $p = 0.05$, whereas protrusion of OPN and pneumatization of ACP were found to be statistically non-significant among this study population (Table 4).

Table 1. Age distribution of patients who participated in the study ($n = 200$).

Age in Years	Minimum	Mean	Maximum
Male	18	51	79
Female	18	37	74
Both	18	43	79

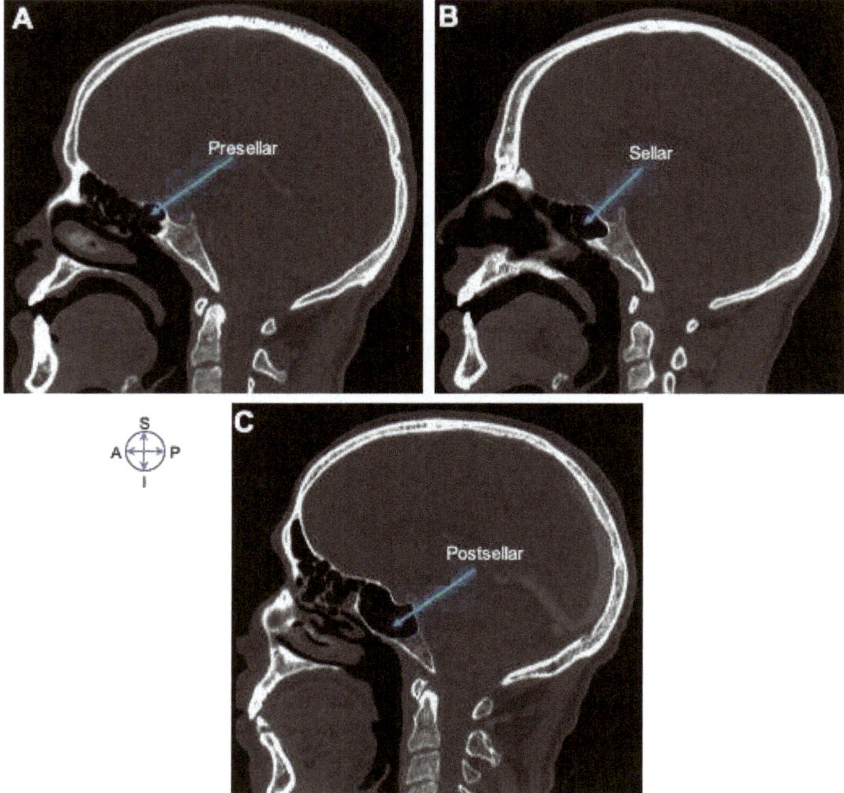

Figure 1. Anatomographic variants of sphenoid sinus. Sagittal CT images showing (**A**) presellar, (**B**) sellar, and (**C**) postsellar types of sphenoid sinus. Note that the sellar type of pneumatization was found to be the most frequent anatomographic variant of sphenoid sinus in the Ethiopian population. Image orientation: S, superior; I, inferior; A, anterior; P, posterior.

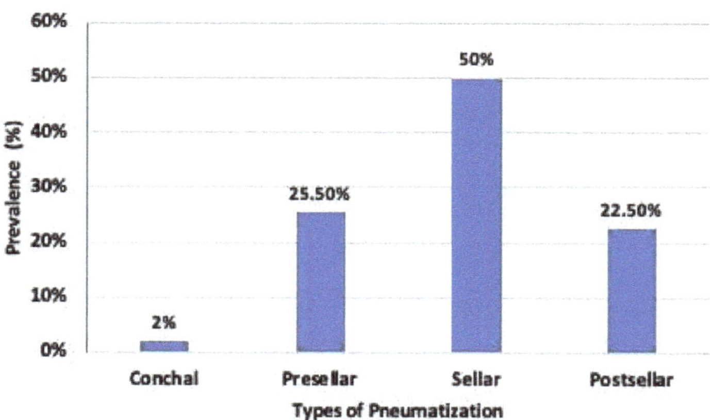

Figure 2. Prevalence of type of sphenoid sinus pneumatization.

Table 2. Incidence of sphenoid sinus pneumatization and septation in the Ethiopian population.

Anatomographic Variants of the Sphenoid Sinus					
Pneumatization		**Septation**			
Type	**Frequency**	**Type**	**Midline**	**Right Side**	**Left Side**
Conchal	4 (2%)	Single complete	76 (38%)	73 (36.5%)	6 (3%)
Presellar	51 (25.5%)	Double complete	1 (0.5%)	19 (9.5%)	0 (0%)
Sellar	100 (50%)	No septa	2(1%)	0 (0%)	0 (0%)
Postsellar	45 (22.5%)				

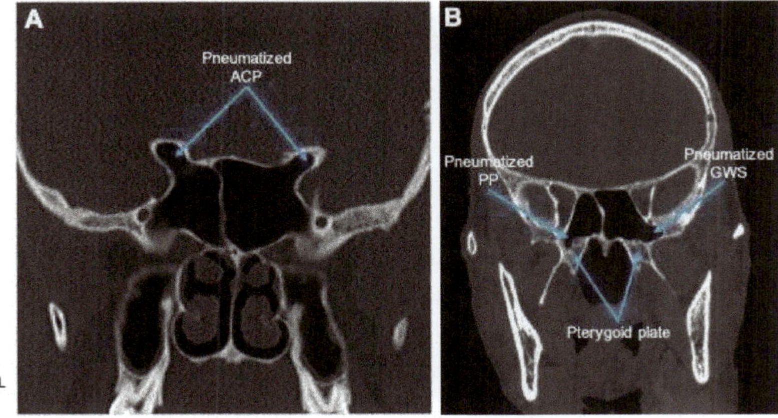

Figure 3. Pneumatization of the sphenoid sinus. Coronal CT images showing (**A**) Pneumatized anterior clinoid process (ACP); (**B**) Pneumatized pterygoid plate (PP) and greater wing of sphenoid (GWS). Note that half of the patient population participating in the study possessed sellar type of sphenoid sinus pneumatization. Image orientation: S, superior; I, inferior; R, right side; L, left side.

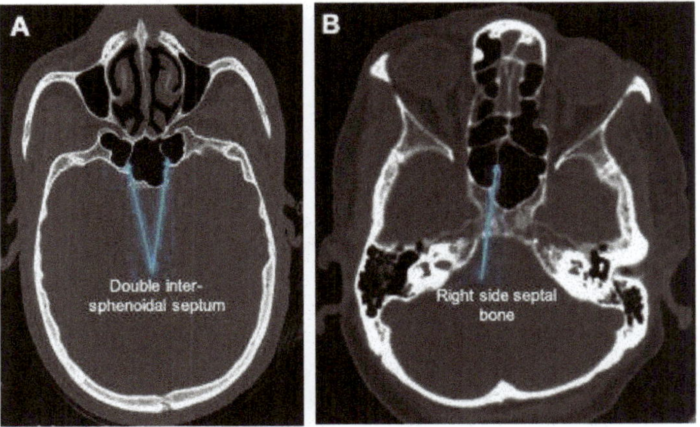

Figure 4. *Cont.*

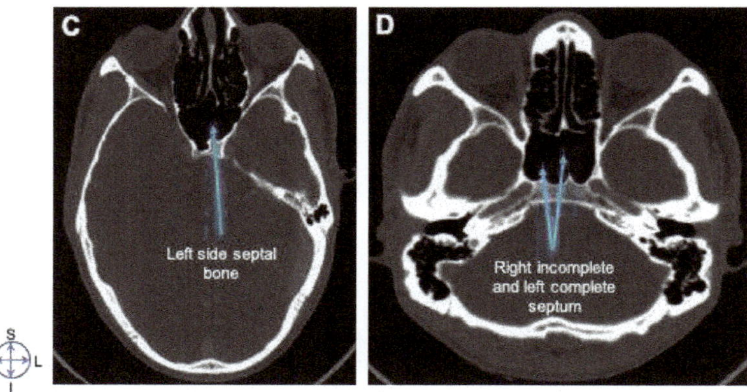

Figure 4. Septation of the sphenoid sinus. Axial CT images showing (**A**) Double inter-sphenoidal septum, (**B**) Right side septal bone, (**C**) Left side septal bone, and (**D**) Right incomplete and left complete septum. Note that almost 90% of the patient population who participated in this study showed single septation, whereas 1% without septation. Image orientation: S, superior; I, inferior; R, right side; L, left side.

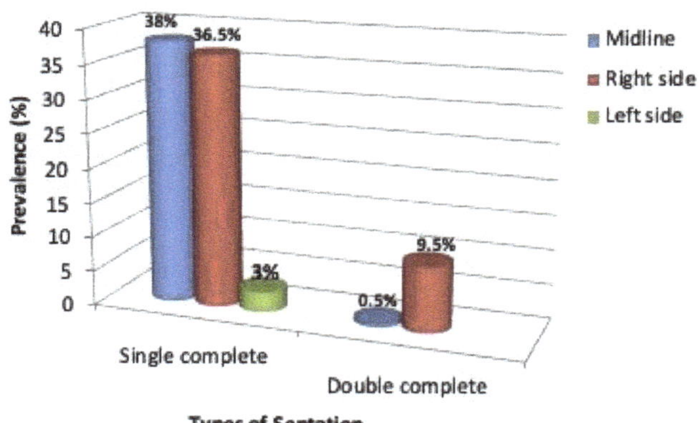

Figure 5. Sphenoid sinus septum in midline, right and left side orientation.

Table 3. Summary of relationship of internal carotid artery, optic nerve and foramen rotundum to the sphenoid sinus.

Structure		Sides			
		Right Side	Left Side	Bilateral	Total
ICA *	Protrusion	12 (32.4%)	9 (24.3%)	16 (43.2%)	37 (18.5%)
	Dehiscence	7 (29.17%)	10 (41.7%)	7 (29.17%)	24 (12%)
OPN *	Protrusion	1 (5.26%)	10 (52.63%)	8 (42.1%)	19 (9.5%)
	Dehiscence	12 (38.7%)	9 (29%)	10 (32.25%)	31 (15.5%)
V2 *	Protrusion	7 (28%)	8 (32%)	10 (40%)	25 (12.5%)
	Dehiscence	4 (16%)	11 (44%)	10 (40%)	25 (12.5%)

* ICA (internal carotid artery); OPN (optic nerve); V2 (foramen rotundum).

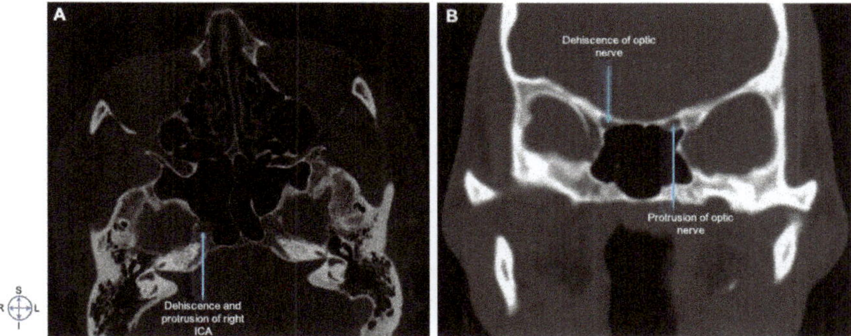

Figure 6. Dehiscence and protrusion of sphenoid sinus in relationship to internal carotid artery (ICA) and optic nerve. Axial CT image (**A**) showing dehiscence and protrusion of right ICA, (**B**) showing dehiscence of optic nerve (R) and protrusion of optic nerve (L). Note the differential protrusion and dehiscence in Ethiopian population. Image orientation: S, superior; I, inferior; R, right side; L, left side.

Table 4. Chi-square Tests.

(df = 1)	Pearson Chi-Square-χ^2	p 95% Confidence Interval
Dehiscence of VDC * pneumatization PP *	7.864	0.049
Protrusion of OPN * pneumatization of ACP *	0.584	0.747
Dehiscence of OPN * pneumatization of ACP *	7.945	0.046

* VDC, Vidian Canal; PP, Pterygoid Plate; OPN, Optic Nerve; ACP, Anterior Clinoid Process. (df = 1) Pearson Chi-square-χ^2 p 95% confidence interval pterygoid plate, optic nerve and anterior clinoid process.

4. Discussion

Anatomical variations of the sphenoid sinus is well documented among many African populations [5,17], however, information is scanty among the Horn of African nations. Among them, Ethiopia is often considered as the cradle of mankind, the land of origins. Understanding anatomographic variants of the sphenoid sinus among these populations is of paramount importance to compare the variability or racial differences among black populations [18]. Paranasal sinus lesions are very common and affect a wide range of population with a variety of etiologies from inflammation to neoplasm [19]. Disease of the sphenoid sinus is difficult to diagnose and treat due to the fact that initial symptoms are vague to recognize. In addition to this, the rarity of sphenoid sinus involvement can be explained by the nonspecific symptoms, the inaccessibility to the sinus through the otorhinolaryngological physical examination, and the low number of diagnoses prior to the advent of more sophisticated technologies such as CT and magnetic resonance imaging [20].

Higher incidence of anatomographical variations of sphenoid sinus can lead to increased risk in terms of injury of important neurovascular and glandular structures [14,21]. Extensive hyperpneumatization of the sphenoid sinus with consecutive pneumatization of ethmoid sinus can lead to injury of the optic nerve [22]. Protrusion of the internal carotid artery into the lumen of the sinuses can also lead to its injury during endoscopic surgical procedures, especially in cases of variation in positions, numbers, and insertions within the sinus septum [23,24].

In this study, we pointed out that compared with presellar and postsellar, sellar was the most frequent anatomographic variant of sphenoid sinus in Ethiopian population (Figures 1 and 2; Table 2). Our finding also indicated that 50% of the Ethiopians participating in the study possess the sellar type of sphenoid sinus pneumatization (Figure 3, Table 2). This finding is in line with the previous study reports of 52.9% [12], but was less than that of 59.4% [9] and 69% [25]. This indicates possible existence of racial anatomographic variants of the sphenoid sinus across different populations. Our findings showed presellar pneumatization in almost a quarter of the 200 study cases (25.5%), where 16.5% patients were on the left side, 6% patients were on the right, and 3% patients were bilateral. The conchal type of pneumatization was the least frequent (2%) in our study, where as in other studies it was 1.8% [2,19], 1.9% [12], and 3% [25], and these differences might be attributed to the variations in sample size selection in various studies. Our findings of pneumatization of ACP 18%, PP 15%, and GWS 16.5% was consistent with previous studies done at "Ovidius" University of Constanta, New Jersey, USA, and Oyo-State, Nigeria [4,19,25,26]. Furthermore, a study done in Constanta Romania found that 33% of the cases showed extensive pneumatization of the pterygoid process; in 10% of the cases, the anterior clinoid processes were pneumatized, forming deep optico-carotid recesses; and in 8% of the cases, there were lateral extensions of pneumatization in the greater wing of the sphenoid bone [25].

Sphenoid sinus septal bone variations are also considered the potential anatomographical variations illustrated in this region [27]. Our study elucidated that almost 90% of Ethiopian population participating in the study had single septation, whereas 1% were without septation. Out of the single septated cases, 38% were midline, 36.5% right side, and 3% on the left side, whereas single incomplete septa and double septa corresponded with each other at 11.5%. However, the absence of septa was observed only in 1% of the study population ((Figures 4 and 5; Table 2)). Regarding the degree of septation, a single inter-sphenoidal septum was observed in 28.1%, and more than one septum in 71.9%. When this finding is compared with other study populations, it is seemingly higher than other studies done in Kerala, India, Turkey, and South Africa populations [7,28–30]. These results were different from other findings, possibly due to the presence of racial or anatomographical variants across different geographical or ancestral origins that might show the evolution of anatomical variations along ethno-geographical or racial diversity. In our study, the absence of septa was found in only 1% of the patients, while it was 2.7% in Nigerians [4,31] and 2.2% in other studies [32–34]. These findings are almost comparable despite the difference in their sample size. Corroborated with our finding, they all point to the rare occurrence of sphenoid sinus without septation in less than 2% across the different studied populations.

Our investigation on the relationship of the sphenoid sinus with ICA, OPN, and V2 are summarized in Figure 6 and Table 3. Protrusion and dehiscence of the sphenoid sinus with ICA was 18.5% and 12%, with OPN of 9.5% and 15.5%, and with V2 of 12.5% and 12.5%, respectively. This entails radiological evaluation that showed an equivalent protrusion and dehiscence of V2, higher protrusion of ICA, and higher dehiscence of OPN among the Ethiopians participated in the study, which is a clinically relevant anatomographic variant to be considered during surgical interventions involving the sphenoid sinus. These findings are also consistent with other findings [23,31].

In summary, the sellar type of sphenoid sinus pneumatization was found to be the most clinically relevant anatomographic variant among the Ethiopian population. Ninety percent of the patient population participating in the study were found to be tomographically single septated. A radiologically equivalent protrusion and dehiscence of the sphenoid sinuses can be considered as anatomographical variations during neurosurgical interventions in the Ethiopian population. The vast majority of this variability is attributed to the extent of sphenoid sinus pneumatization, varying number and position of septae, and the anatomical relationship with surrounding neurovascular and glandular structures visible during imaging modalities. To the best of our knowledge, this is the first diagnostic report in Eastern Africa incorporating both anatomical and radiological data. We also recommend further investigation in the populations of common ancestral origins.

Author Contributions: T.K.D. carried out the study. A.M.Z., A.T.W., and S.T.D. supervised the research. T.D.W. interpreted the radiological images together with senior residents. J.M.G. designed the manuscript writing for correspondence. All authors contributed ideas for this study. Further multiple revisions and editing for publication was done by A.T.W., A.M.Z., and J.M.G. All authors have read and agreed to the published version of the manuscript.

Funding: This research received no external funding.

Acknowledgments: We would like to thank Addis Ababa University, Tikur Anbessa Specialized Teaching Hospital, for providing assistance. We are also very grateful to radiology residents, who assisted us during data collection.

Conflicts of Interest: The authors declare that there are no conflicts of interests.

References

1. Jaworek-Troć, J.; Zarzecki, M.; Zamojska, I.; Iwanaga, J.; Przybycień, W.; Mazur, M.; Chrzan, R.; Walocha, J.A. The dimensions of the sphenoid sinuses—Evaluation before the functional endoscopic sinus surgery. *Folia Morphol.* **2020**. [CrossRef] [PubMed]
2. Anusha, B.; Baharudin, A.; Philip, R.; Harvinder, S.; Shaffie, B.M. Anatomical variations of the sphenoid sinus and its adjacent structures: A review of existing literature. *Surg. Radiol. Anat.* **2014**, *36*, 419–427. [CrossRef] [PubMed]
3. Ozturan, O.; Yenigun, A.; Degirmenci, N.; Aksoy, F.; Veyseller, B. Co-existence of the Onodi cell with the variation of perisphenoidal structures. *Eur. Arch. Otorhinolaryngol.* **2013**, *270*, 2057–2063. [CrossRef] [PubMed]
4. Fasunla, A.J.; Ameye, S.A.; Adebola, O.S.; Ogbole, G.; Adeleye, A.O.; Adekanmi, A.J. Anatomical variations of the sphenoid sinus and nearby neurovascular structures seen on computed tomography of black Africans. *East Cent. Afr. J. Surg.* **2012**, *17*, 57–64. [CrossRef]
5. Awadalla, A.M.; Hussein, Y.; ELKammash, T.H. Anatomical and radiological parameters of the sphenoid sinus among Egyptians and its impact on sellar region surgery. *Egypt. J. Neurosurg.* **2015**, *30*, 1–12.
6. Štoković, N.; Trkulja, V.; Dumić-Čule, I.; Čuković-Bagić, I.; Lauc, T.; Vukičević, S.; Grgurević, L. Sphenoid sinus types, dimensions and relationship with surrounding structures. *Ann. Anat.* **2016**, *203*, 69–76. [CrossRef]
7. Sevinc, O.; Is, M.; Barut, C.; Erdogan, A. Anatomic Variations of Sphenoid Sinus Pneumatization in a Sample of Turkish Population: MRI Study. *Int. J. Morphol.* **2014**, *32*, 1140–1143. [CrossRef]
8. Budu, V.; Mogoantă, C.A.; Fănuţă, B.; Bulescu, I. The anatomical relations of the sphenoid sinus and their implications in sphenoid endoscopic surgery. *Rom. J. Morphol. Embryol.* **2013**, *54*, 13–16.
9. Seddighi, A.; Seddighi, A.S.; Mellati, O.; Ghorbani, J.; Raad, N.; Soleimani, M.M. Sphenoid Sinus: Anatomic Variations and Their Importance in Trans-sphenoid Surgery. *Int. Clin. Neurosci. J.* **2014**, *1*, 31–34.
10. Craiu, C.; Sandulescu, M.; Rusu, M.C. Variations of sphenoid pneumatization: A CBCT study. *Rom. J. Rhinol.* **2015**, *5*, 107–113. [CrossRef]
11. Pérez-Piñas, I.; Sabaté, J.; Carmona, A.; Catalina-Herrera, C.J.; Jiménez-Castellanos, J. Anatomical variations in the human paranasal sinus region studied by CT. *J. Anat.* **2000**, *197*, 221–227. [CrossRef] [PubMed]
12. Kayalioglu, G.; Erturk, M.; Varol, T. Variations in sphenoid sinus anatomy with special emphasis on pneumatization and endoscopic anatomic distances. *Neurosciences (Riyadh)* **2005**, *10*, 79–84. [PubMed]
13. Chougule, M.S.; Dixit, D. A Cross-Sectional Study of Sphenoid Sinus through Gross and Endoscopic Dissection in North Karnataka, India. *J. Clin. Diagn. Res.* **2014**, *8*, AC01–AC5. [CrossRef] [PubMed]
14. Kantarci, M.; Karasen, R.M.; Alper, F.; Onbas, O.; Okur, A.; Karaman, A. Remarkable anatomic variations in paranasal sinus region and their clinical importance. *Eur. J. Radiol.* **2004**, *50*, 296–302. [CrossRef]
15. Raudaschl, P.F.; Zaffino, P.; Sharp, G.C.; Spadea, M.F.; Chen, A.; Dawant, B.M.; Albrecht, T.; Gass, T.; Langguth, C.; Lüthi, M.; et al. Evaluation of segmentation methods on head and neck CT: Auto-segmentation challenge 2015. *Med. Phys.* **2017**, *44*, 2020–2036. [CrossRef]
16. Lokwani, M.S.; Patidar, J.; Parihar, V. Anatomical variations of sphenoid sinus on multi-detector computed tomography and its usefulness in trans-sphenoidal endoscopic skull base surgery. *Int. J. Res. Med. Sci.* **2018**, *6*, 3063–3071. [CrossRef]
17. Kajoak, S.A.; Ayad, C.E.; Najmeldeen, M.; Abdalla, E.A. Computerized tomography morphometric analysis of the sphenoid sinus and related structures in Sudanese population. *Glob. Adv. Res. J. Med. Med. Sci.* **2014**, *3*, 160–167.
18. Hewaidi, G.; Omami, G. Anatomic Variation of Sphenoid Sinus and Related Structures in Libyan Population: CT Scan Study. *Libyan J. Med.* **2008**, *3*, 128–133. [CrossRef]

19. Tomovic, S.; Esmaeili, A.; Chan, N.J.; Shukla, P.A.; Choudhry, O.J.; Liu, J.K.; Eloy, J.A. High-resolution computed tomography analysis of variations of the sphenoid sinus. *J. Neurol. Surg. Part. B Skull Base* **2013**, *74*, 82–90. [CrossRef]
20. Hamid, O.; El Fiky, L.; Hassan, O.; Kotb, A.; El Fiky, S. Anatomic Variations of the Sphenoid Sinus and Their Impact on Trans-sphenoid Pituitary Surgery. *Skull Base* **2008**, *18*, 9–15. [CrossRef]
21. Shpilberg, K.A.; Daniel, S.C.; Doshi, A.H.; Lawson, W.; Som, P.M. CT of Anatomic Variants of the Paranasal Sinuses and Nasal Cavity: Poor Correlation with Radiologically Significant Rhinosinusitis but Importance in Surgical Planning. *AJR Am. J. Roentgenol.* **2015**, *204*, 1255–1260. [CrossRef]
22. Heskova, G.; Mellova, Y.; Holomanova, A.; Vybohova, D.; Kunertova, L.; Marcekova, M.; Mello, M. Assessment of the relation of the optic nerve to the posterior ethmoid and sphenoid sinuses by computed tomography. *Biomed. Pap. Med. Fac. Univ. Palacky Olomouc. Czech. Repub.* **2009**, *153*, 149–152. [CrossRef] [PubMed]
23. Jaworek-Troć, J.; Zarzecki, M.; Bonczar, A.; Kaythampillai, L.N.; Rutowicz, B.; Mazur, M.; Urbaniak, J.; Przybycień, W.; Piątek-Koziej, K.; Kuniewicz, M.; et al. Sphenoid bone and its sinus-anatomo-clinical review of the literature including application to FESS. *Folia Med. Cracov.* **2019**, *59*, 45–59. [PubMed]
24. Raseman, J.; Guryildirim, M.; Beer-Furlan, A.; Jhaveri, M.; Tajudeen, B.A.; Byrne, R.W.; Batra, P.S. Preoperative Computed Tomography Imaging of the Sphenoid Sinus: Striving Towards Safe Transsphenoidal Surgery. *J. Neurol. Surg. B Skull Base* **2020**, *81*, 251–262. [CrossRef] [PubMed]
25. Lupascu, M.; Comsa, G.I.; Zainea, V. Anatomical variations of the sphenoid sinus-a study of 200 cases. *ARS Medica Tomitana* **2014**, *20*, 57–62. [CrossRef]
26. Rysz, M.; Bakoń, L. Maxillary sinus anatomy variation and nasal cavity width: Structural computed tomography imaging. *Folia Morphol. (Warsz)* **2009**, *68*, 260–264. [PubMed]
27. Hiremath, S.B.; Gautam, A.A.; Sheeja, K.; Benjamin, G. Assessment of variations in sphenoid sinus pneumatization in Indian population: A multidetector computed tomography study. *Indian J. Radiol. Imaging* **2018**, *28*, 273–279. [CrossRef]
28. Ngubane, N.P.; Lazarus, L.; Rennie, C.O.; Satyapal, K.S. The Septation of the Sphenoidal Air Sinus. A Cadaveric Study. *Int. J. Morphol.* **2018**, *36*, 1413–1422. [CrossRef]
29. Jaworek-Troć, J.; Zarzecki, M.; Mróz, I.; Troć, P.; Chrzan, R.; Zawiliński, J.; Walocha, J.; Urbanik, A. The total number of septa and antra in the sphenoid sinuses-evaluation before the FESS. *Folia Med. Cracov.* **2018**, *58*, 67–81. [CrossRef]
30. Guga Priya, T.S.; Kumar, N.V.; Guru, A.T.; NalinaKumari, S.D. An Anatomo-Imagistic Study of Intersphenoidal Sinus Septum. *Int. J. Anat. Res.* **2016**, *4*, 2015–2020. [CrossRef]
31. Akanni, D.; Souza, C.; Savi de Tove, K.M.; N'zi, K.; Yèkpè, P.; Biaou, O.; Boco, V. Sphenoid Sinuses Pneumatization and Association with the Protrusion of Surrounding Neurovascular Structures amongst Beninese. *Open J. Radiol.* **2018**, *8*, 209–216. [CrossRef]
32. Riza, D.; Erkan, K.; Fatih, S.; Mehmet, A.; Ahmet, K.; Engin, S.; Haflmet, Y.; Abdülkadir, E. Radiological evaluation of septal bone variations in the sphenoid sinus. *J. Med. Updates* **2014**, *4*, 6–10. [CrossRef]
33. Hengerer, A.S. Surgical anatomy of the paranasal sinuses. *Ear Nose Throat J.* **1984**, *63*, 137–143.
34. Bedawi, K.; Madani, G.A.; Seddeg, Y. The Radiological Study of Onodi Cells Among Adult Sudanese Subjects. *OSR-JDMS* **2017**, *16*, 106–109. [CrossRef]

Publisher's Note: MDPI stays neutral with regard to jurisdictional claims in published maps and institutional affiliations.

Article

Longitudinal Observation of Changes in the Ankle Alignment and Tibiofibular Relationships in Hereditary Multiple Exostoses

Jae Hoo Lee [1], Chasanal Mohan Rathod [2], Hoon Park [3], Hye Sun Lee [4], Isaac Rhee [5] and Hyun Woo Kim [6,*]

1. Department of Orthopaedic Surgery, Ilsan Paik Hospital, Inje University College of Medicine, Goyang 10380, Korea; jaehoolee@gmail.com
2. Division of Pediatric Orthopaedic Surgery, SRCC NH Children's Hospital, Mumbai 400034, India; chasanal@gmail.com
3. Department of Orthopaedic Surgery, Gangnam Severance Hospital, Yonsei University College of Medicine, Seoul 06273, Korea; hoondeng@yuhs.ac
4. Biostatistics Collaboration Unit, Yonsei University College of Medicine, Seoul 03722, Korea; hslee1@yuhs.ac
5. Medical Course, University of Melbourne Melbourne Medical School, Melbourne 3010, Australia; isaac.rhee@gmail.com
6. Division of Pediatric Orthopaedic Surgery, Severance Children's Hospital, Yonsei University College of Medicine, Seoul 03722, Korea

* Correspondence: pedhkim@yuhs.ac; Tel.: +82-2-2228-2180

Received: 6 September 2020; Accepted: 25 September 2020; Published: 25 September 2020

Abstract: The longitudinal changes in the tibiofibular relationship as the ankle valgus deformity progresses in patients with hereditary multiple exostoses (HME) are not well-known. We investigated the longitudinal changes and associating factors in the tibiofibular relationship during the growing period. A total of 33 patients (63 legs) with HME underwent two or more standing full-length anteroposterior radiographs. Based on the change in ankle alignments, thirty-five patients with an increase in tibiotalar angle were grouped into group V, and 28 patients with a decreased angle into group N. In terms of the change in radiographic parameters, significant differences were noted in the tibial length, the fibular/tibial ratio, and the proximal and distal epiphyseal gap. However, age, sex, initial ankle alignment, location of osteochondroma, and presence of tibiofibular synostosis did not affect the tibiofibular alignment. The tibial growth was relatively greater than the fibular growth and was accompanied by significant relative fibular shortening in the proximal and distal portions. In pediatric patients with HME, age, sex, initial ankle alignment, location of the osteochondroma, and synostosis did not predict the progression of the ankle valgus deformity. However, when valgus angulation progressed, relative fibular shortening was observed as the tibia grew significantly in comparison to the fibula.

Keywords: hereditary multiple exostoses; tibia; fibula; ankle; valgus

1. Introduction

Valgus deformity of the ankle is a common manifestation of hereditary multiple exostoses (HME), affecting approximately half of the patients [1–4]. It may result in pain, restricted range of motion, and gait disturbances [5]. In a study investigating the natural history of untreated ankle joints in HME patients, 14 out of 75 ankles (19%) with an average tibiotalar tilt of valgus 9° showed early radiographic signs of osteoarthritis [6]. Appropriate surgical correction for excessive tibiotalar tilting has been recommended to preserve ankle function and prevent early arthritic changes in adulthood [7–9].

Past studies have suggested that the ankle valgus deformity is caused by a disproportionate shortening of the fibula relative to the tibia [2,4,6,10]. Compared to valgus deformity of the knee, which is primarily caused by changes in the distal femur and proximal tibia, ankle valgus deformity is directly affected by the disparity between the distal tibial and fibular length [1,6]. For this reason, predicting fibular shortening in HME patients may guide the indications for appropriate surgical treatment of ankle valgus.

Characteristic factors that have been associated with higher rates of ankle valgus deformities include an ankle valgus over 10° in males and exostoses on the proximal and distal tibia or fibula [10,11]. However, there is a lack of research analyzing the tibiofibular relationship, especially longitudinal observation studies investigating tibial and fibular growth. Furthermore, the literature is sparse in identifying factors related to the progression of ankle valgus and relative fibular shortening during growth periods.

The purpose of this study was to: (1) analyze longitudinal changes in the tibiofibular relationship of HME patients during their growing period; (2) identify factors associated with the progression of ankle valgus deformities; and (3) determine the effects of predisposing factors on the growth of the fibula and tibia over time.

2. Materials and Methods

2.1. Patient Recruitment

This retrospective study was approval by the institutional review board of the Severance Hospital, Seoul, Korea (28 November 2013; 4-2013-0700). Medical records and plain radiographs of patients who were skeletally immature at their first visit and underwent two or more full-length standing anteroposterior (AP) radiographs of the lower extremities during their follow-up were reviewed. Patients with a previous surgical history of the lower extremities and a limb-length discrepancy of over 2.5 cm, measured in any of the radiographs, were excluded. Among 53 patients who were initially enrolled, a total of 20 patients were excluded due to surgical history of their lower extremities (16 patients) and limb-length discrepancy of over 2.5 cm (4 patients). A total of 265 full-length radiographs of 63 affected legs in 33 patients were finally included in the study. The cases were divided into two groups of Group V and Group N, according to whether the tibiotalar angle increased into a more valgus position (Group V), or decreased into a more varus position (Group N) at the final evaluation when compared to the initial visit.

2.2. Radiographic Measurements

Full-length AP radiographs were taken in the standardized manner with the patients in a bipedal stance with equal weight bearing of both feet [12]. All radiographic parameters were measured using Picture Archiving and Communication Systems (PACS) (Centricity PACS 2.0; GE Medical Systems Information Technologies, Milwaukee, WI, USA). Radiographic parameters [1,2,4,10] were as follows: lateral distal femoral angles (LDFA), medial proximal tibial angles (MPTA), proximal and distal epiphyseal and physeal gaps in the tibia and fibula, fibula and tibia length, fibula/tibia length ratio, tibiotalar angles, locations of osteochondromas in the tibia or fibula, and the presence of tibiofibular synostosis.

Anatomical LDFA was defined as the lateral angle between the longitudinal axis of the femoral shaft and a line across the surface of the distal femoral epiphysis. Mechanical MPTA was defined as the medial angle between the longitudinal axis of the tibial shaft and a line across the tibial plateaus. The fibular length was measured as the straight-line distance from the tip of the proximal epiphysis to the tip of the lateral malleolus; the tibial length was measured as the straight-line distance from the tips of the tibial eminences to the tibial plafond. We also measured the distance between the tips of the tibial and fibular epiphyses, respectively. Proximally, due to the complex anatomy of the proximal tibial epiphysis, the distance was defined as the gap between a parallel line past the apex of the proximal fibula and a parallel line through the midpoint between the line along the apex of the tibial spine and a line on the bottom-most portion of the condyle. Additionally, a parallel line through the midpoints between a line passing the top of the physis and a line crossing the bottom of physis were set as a reference line in order to measure the distance between the proximal and distal fibular and tibial physes (Figure 1A). The tibiotalar angle was defined as the medial angle between a perpendicular line to the axis of the tibia and the extended line that touches the articular surface of the talus (Figure 1B) [6]. In order to categorize the initial ankle alignment, the cases were divided into three positions: neutral (tibiotalar angle = 0–5°), valgus (tibiotalar angle > 5°), and varus (tibiotalar angle < 0°) [13].

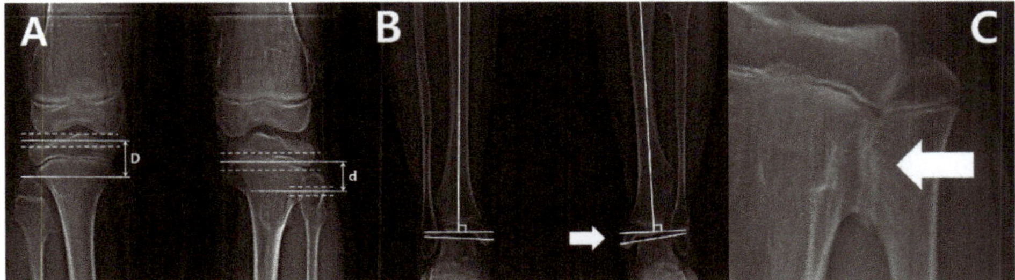

Figure 1. (**A**) The distance of proximal epiphyses (D) and physes (d) between tibia and fibula. The distance of proximal epiphyses was defined as the gap between the parallel line past the apex of the proximal fibula and the parallel line past the center of the distance between the line across the apex of the tibial spine and the most bottom of the condyle. The distances of proximal physes was defined as the gap between the parallel line passing through the midpoint of the distance between the line across the top of physis; the line crossing the bottom of physis was set as the reference line. (**B**) The measurement of the ankle joint. The tibiotalar angle was defined as the extension line of the tibial anatomical axis and a line perpendicular to the extension line that touches the dome of the talus. The left lower extremity indicated by the white arrow demonstrates a decreased distal tibiofibular distance compared to the right side, and a prominent ankle valgus deformity is also observed. (**C**) The tibiofibular synostosis. The white arrow indicates a definite bridge of an exostosis lesion connecting the proximal fibula and tibia without the overlapping cortex.

The presence of tibiofibular synostosis was determined when definite bridging by an exostosis lesion of the tibia and fibula, without overlapping cortices, was identified (Figure 1C). The location of osteochondromas were determined by dividing the entire length of the fibula and tibia, including both ends of the proximal and distal epiphysis, into two equal parts [10].

Reliability testing of radiographic measurements was performed. Two orthopedic surgeons independently measured the radiographic parameters, and inter-observer reliability of the two surgeons was determined using the intraclass correlation coefficient (ICC). Both surgeons also repeated measurements of the same subject at 2-week intervals to assure intra-observer reliability.

2.3. Statistical Analyses

The intra- and inter-observer reliabilities of the radiographic measurements were analyzed using the ICC. ICC was calculated using a two-way mixed effect for an absolute agreement between both values of each observer's measurements. An ICC value of 0.75 or greater was considered to reflect excellent reliability [14]. All statistical analyses were performed using SAS version 9.4 (SAS Institute, Cary, NC, USA) and R statistics 3.6.2. p value < 0.05 was indicated as statistically significant.

Numerical variables were presented as mean ± standard deviation and frequency (percentage). Independent two-sample t tests were used to compare patient demographics and radiographic measurements between the groups of N and V. Categorical variables such as sex, the presence of tibiofibular synostosis, and location of osteochondroma were compared using a chi-square test (Fisher's exact test). Mann–Whitney U-test was used to compare the follow-up duration between the groups. The linear mixed model, for a repeated-measures random intercept model, was used for LDFA, MPTA, proximal and distal epiphyseal and physeal gaps between the tibia and fibula, tibia and fibula length, fibula/tibia length ratio, and tibiotalar angle. The changes in the fibular length, tibial length, fibula/tibia ratio, and tibiotalar angle according to time were considered as covariate x time interactions. The fixed effects of associated factors on tibial and fibular lengths, and the tibiotalar angle per month which can affect the growth of lower extremities including patient's age, sex, synostosis and initial ankle alignment were analyzed.

3. Results

3.1. Inter-and Intra-Observer Reliability

The interobserver reliability for the radiographic measurements of the lower-limb alignment were found to be excellent (ICC; LDFA, 0.973; MPTA, 0.947; fibular length, 0.999; tibia; length, 0.999; proximal physis gap, 0.989; proximal epiphysis gap, 0.979; distal physis gap, 0.964; distal epiphysis, 0.953; tibiofibular angle, 0.993). The intraobserver reliability for the observer's repeated measurements also resulted in excellent results (ICC; LDFA, 0.946; MPTA, 0.968; fibular length, 0.999; tibia; length, 0.999; proximal physis gap, 0.995; proximal epiphysis gap, 0.990; distal physis gap, 0.985; distal epiphysis, 0.968; tibiofibular angle, 0.999).

3.2. Demographic Distribution

The demographic data of patients at their first visit are summarized in Table 1.

Table 1. Summary of the patients.

	Group N (n = 28)	Group V (n = 35)	p Value
Numbers of affected leg (right:left)	15:13	18:17	
Age at first visit (months)	77.6 ± 40.8	82.8 ± 33.9	0.586
Sex (male:female)	14:14	17:18	0.910
Duration of follow-up (months)	27.1 ± 14.5	40.5 ± 19.8	0.004 *
Numbers of radiologic follow-up	3.4 ± 1.6	4.8 ± 2.1	0.003 *
Initial ankle alignment †			0.964
Neutral (tibiotalar angle 0°–5°)	14 (51.9%)	17 (48.6%)	
Valgus (tibiotalar angle > 5°)	7 (26.0%)	10 (28.6%)	
Varus (tibiotalar angle < 0°)	6 (22.2%)	8 (22.9%)	
Location of osteochondroma			
Proximal tibia	17 (63.0%)	26 (74.3%)	0.250
Distal tibia	12 (44.4%)	21 (60.0%)	0.176
Proximal fibula	18 (66.7%)	27 (77.1%)	0.262
Distal fibula	12 (44.4%)	15 (42.9%)	>0.999
Tibiofibular synostosis			0.906
None	10 (37.0%)	12 (34.3%)	
Presence	18 (66.7%)	23 (65.7%)	

Values are presented as mean ± standard deviation or as numbers only. * Statistical significance was noted. † The valgus angle is expressed as a positive value, and the varus angle as a negative value.

Age at the initial visit did not differ between the groups; however, the follow-up period for group V was significantly longer than that of group N (group N, 27.1 ± 14.5 months; group V, 40.5 ± 19.8 months, p = 0.004). Group V also had a significantly greater number of radiographs taken for each leg than group N (p = 0.003). The distribution of initial ankle alignment in both groups was similar. Neutral alignment was most commonly seen, followed by valgus and varus alignments. There were no significant differences in the location of exostosis (proximal tibia, p = 0.250; distal tibia, p = 0.176; proximal fibula, p = 0.262, and distal fibula, p > 0.999, respectively) and the presence of tibiofibular synostosis (p = 0.906) between the groups. The mean tibiofibular alignments at the initial and last evaluation are displayed in Table 2. Based on the initial evaluation, no significant differences in any of the parameters were noted between the groups. However, the final evaluation revealed significant differences between the groups in all parameters, except for the distal physes gap.

Table 2. Measurements of lower limb alignment at initial visit and last follow-up.

	Group N	Group V	*p* Value
Initial visit			
Lateral distal femoral angle (°)	89.1 ± 4.2	87.7 ± 2.5	0.111
Medial proximal tibial angle (°)	89.3 ± 3.1	88.9 ± 2.3	0.567
Length of fibula (mm)	233.4 ± 59.0	240.8 ± 47.2	0.587
Length of tibia (mm)	239.9 ± 61.1	246.9 ± 48.4	0.620
Fibula/tibia ratio	0.97 ± 0.04	0.98 ± 0.34	0.834
Gap of proximal physis (mm)	11.9 ± 5.8	14.8 ± 6.0	0.058
Gap of proximal epiphysis (mm)	15.0 ± 5.0	15.9 ± 4.5	0.495
Gap of distal physis (mm)	6.7 ± 3.8	6.6 ± 3.6	0.905
Gap of distal epiphysis (mm)	8.6 ± 4.0	9.7 ± 4.3	0.329
Tibiotalar angle (°) †	2.6 ± 7.0	2.9 ± 5.5	0.876
Last follow-up			
Lateral distal femoral angle (°)	88.4 ± 3.0	86.5 ± 2.3	0.007 *
Medial proximal tibial angle (°)	90.5 ± 3.3	88.1 ± 2.6	0.003 *
Length of fibula (mm)	260.9 ± 47.7	284.9 ± 42.0	0.040 *
Length of tibia (mm)	265.9 ± 50.2	296.9 ± 43.9	0.012 *
Fibula/tibia ratio	0.98 ± 0.03	0.96 ± 0.03	0.005 *
Gap of proximal physis (mm)	13.5 ± 5.8	19.2 ± 5.5	<0.001 *
Gap of proximal epiphysis (mm)	14.4 ± 4.9	19.1 ± 4.8	<0.001 *
Gap of distal physis (mm)	6.3 ± 3.6	5.2 ± 3.9	0.279
Gap of distal epiphysis (mm)	10.1 ± 3.7	6.0 ± 4.9	0.001 *
Tibiotalar angle (°) †	1.6 ± 7.0	7.5 ± 6.2	0.001 *

Values are presented as mean ± standard deviation or as numbers only.* Statistical significance was noted. † The valgus angle is expressed as a positive value, and the varus angle as a negative value.

3.3. *Change in Alignment During the Observation Period*

The estimated slopes of change between the initial and final measurements were analyzed using a linear mixed model. Group N had a significant constant increase in the proximal physes gap (0.062, $p < 0.001$) and the fibular (0.977, $p < 0.001$) and tibial (0.941; $p < 0.001$) lengths. Group V had a significant increase in the tibia and fibular length, the proximal physes and epiphyses gap, and the tibiotalar angle (Figures 2 and 3), but a decrease in the LDFA, the fibula/tibia ratio, and the distal physes and epiphyses gap. In terms of the differences between the slopes of both groups, significant differences were identified in the tibial length, the fibular/tibial length ratio, the proximal and distal epiphyses gap, and the tibiotalar angle (Table 3).

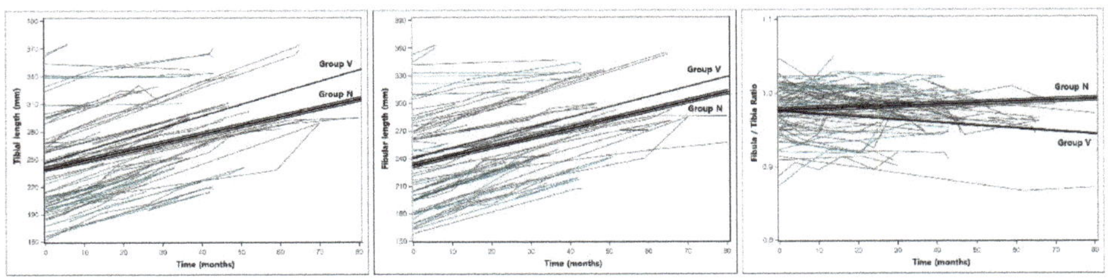

Figure 2. The spaghetti plot and estimated slope in the tibial, fibular length and fibula/tibia ratio of groups N and V. The solid line indicates the estimated slope of group V. and triple lines indicate the estimated slope of group N. No significant difference was noted in fibular length, and there was a significant difference in slope between group N and group V in tibial length and fibula/tibia ratio.

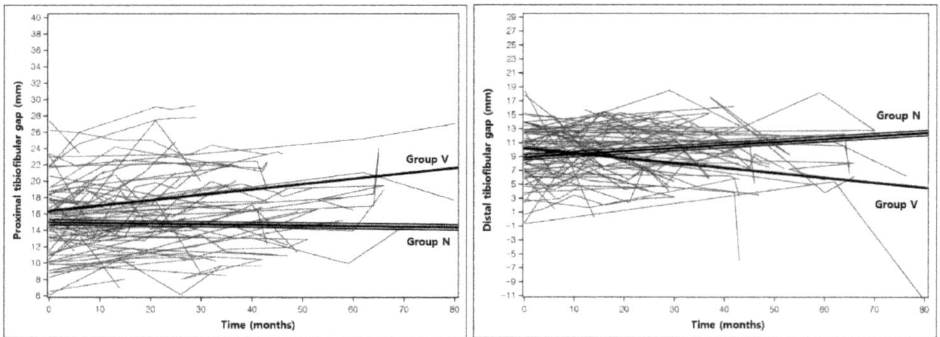

Figure 3. The spaghetti plot and estimated slopes in proximal and distal tibiofibular gaps of groups N and V. The solid line indicates the estimated slope of group V; and triple lines indicate the estimated slope of group N. Progressive fibular shortening with a significant increase in the proximal gap and a decrease in the distal tibiofibular gap were identified over time.

Table 3. Estimated slope of change between the initial and final measurements.

	Estimated Slope (SE)					
	Group N	*p* Value	Group V	*p* Value	Difference between Group N Versus. Group V	*p* Value
Lateral distal femoral angle (°)	−0.028 (0.0144)	0.052	−0.0266 (0.0081)	0.001 *	0.0015 (0.0165)	0.929
Medial proximal tibial angle (°)	0.0256 (0.0148)	0.087	−0.0026 (0.0084)	0.753	−0.0282 (0.017)	0.100
Length of fibula (mm)	0.9765 (0.0612)	<0.001 *	1.0968 (0.0344)	<0.001 *	0.1203 (0.0702)	0.088
Length of tibia (mm)	0.9407 (0.0622)	<0.001 *	1.2483 (0.035)	<0.001 *	0.3076 (0.0713)	<0.001 *
Fibula/tibia ratio	0.0002 (0.0001)	0.066	−0.0004 (0.0001)	<0.001 *	−0.0006 (0.0001)	<0.001 *
Gap of proximal physis (mm)	0.0623 (0.0166)	<0.001 *	0.0978 (0.0093)	<0.001 *	0.0354 (0.019)	0.064
Gap of proximal epiphysis (mm)	−0.0079 (0.0167)	0.637	0.0667 (0.0094)	<0.001 *	0.0746 (0.0192)	<0.001 *
Gap of distal physis (mm)	−0.0084 (0.0145)	0.563	−0.0177 (0.0081)	0.031 *	−0.0093 (0.0166)	0.574
Gap of distal epiphysis (mm)	0.0425 (0.0219)	0.054	−0.0702 (0.0124)	<0.001 *	−0.1126 (0.0251)	<0.001 *
Tibiotalar angle (°) †	−0.0231 (0.0171)	0.180	0.0972 (0.0097)	<0.001 *	0.1202 (0.0197)	<0.001 *

Values are presented as mean (SE, standard error) or as numbers only. * Statistical significance was noted. † The valgus angle is expressed as a positive value, and the varus angle as a negative value.

3.4. *Effects of Characteristic Factors*

The fixed effects of the patient's age, sex, synostosis and initial ankle alignment on subsequent tibial and fibular lengths, and the tibiotalar angle, are listed in Table 4. The amount of change per a month in the fibular and tibial lengths and the ratio of the fibular and tibial lengths decreased significantly (Figure 4). Sex differences were found to not affect the changes in the tibiofibular length ratio or tibiotalar angle. In the presence of synostosis, the rate of growth per month of the tibia and fibula decreased significantly (tibia, −0.125 mm/month; fibula, −0.138 mm/month); however, it did not affect the changes in the ratio of the fibular and tibia lengths or tibiotalar angle. When the initial ankle alignment was valgus, the ratio of the fibular and tibial length increased by 0.0005 per month ($p = 0.004$). For ankles with an initial varus alignment, significant increases in the tibiotalar angle were demonstrated (0.044 mm per month, $p = 0.034$) (Figure 5).

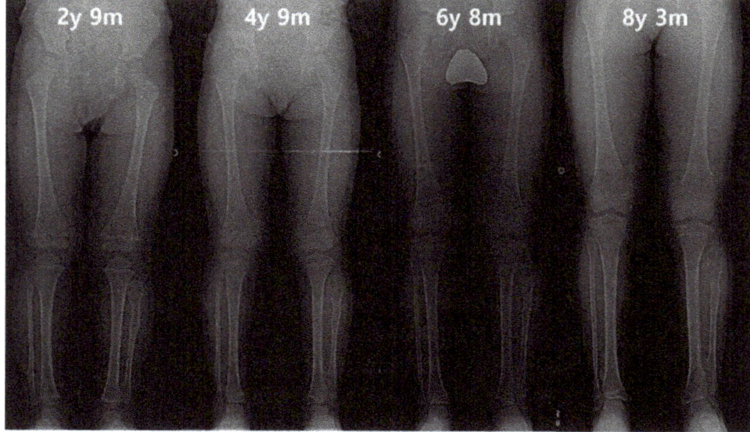

Figure 4. The serial radiographs of a male patient demonstrated progression of valgus deformity on the left ankle. Despite exostoses in bilateral legs, valgus angulation of the left ankle accompanied by shortening of the fibula was observed, while the neutral alignment was maintained on the right ankle. In the last measurement, the left tibial length was also about 9 mm shorter than the right tibia.

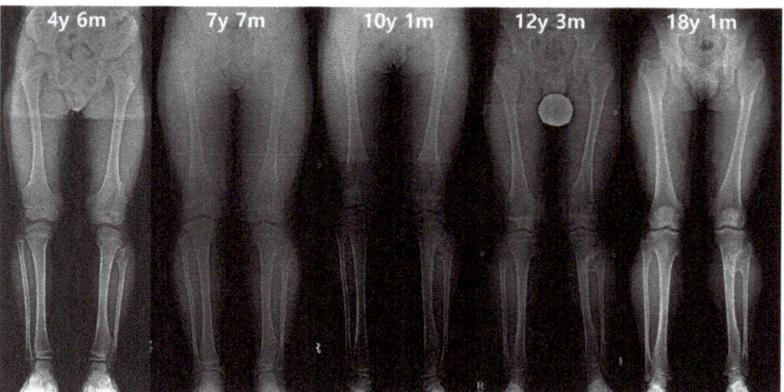

Figure 5. A male patient had prominent exostosis on the left proximal fibula and an increased gap of the proximal tibia and fibula at initial evaluation. Although the right lower extremity initially maintained a normal tibiofibular relationship, after the growth was completed, the right ankle joint developed valgus deformity to be equivalent to the left ankle. Finally, the leg length discrepancy was approximately 12 mm due to the 10 mm shortened length of the left tibia.

Table 4. Interactive effects between the parameters at initial visit and the tibiofibular alignment.

	Fibular Length		Tibial Length		Fibula/Tibia Ratio		Tibiotalar Angle	
	Effect (SE)	***p* Value**	**Effect (SE)**	***p* Value**	**Effect (SE)**	***p* Value**	**Effect (SE)**	***p* Value**
Age	−0.005 (0.001)	<0.001 *	−0.004 (0.001)	<0.001 *	−0.000006 (0.0000018)	0.001 *	0.0001 (0.0003)	0.666
Sex								
Male	Ref. (0) †		Ref. (0) †		Ref. (0) †		Ref. (0) †	
Female	−0.108 (0.058)	0.062	−0.102 (0.063)	0.105	0.000003 (0.00012)	0.979	−0.017 (0.019)	0.360
Synostosis								
None	Ref. (0) †		Ref. (0) †		Ref. (0) †		Ref. (0) †	
Presence	−0.125 (0.058)	0.031 *	−0.138 (0.062)	0.028 *	0.0001 (0.0001)	0.492	−0.013 (0.019)	0.487
Initial ankle alignment								
Neutral (0°- valgus 5°)	Ref. (0) †		Ref. (0) †		Ref. (0) †		Ref. (0) †	
Valgus (> valgus 5°)	−0.032 (0.077)	0.673	−0.114 (0.083)	0.170	0.0005 (0.0002)	0.004 *	0.027 (0.025)	0.278
Varus (< 0°)	0.080 (0.063)	0.204	0.024 (0.068)	0.726	0.0002 (0.0001)	0.097	0.044 (0.021)	0.034 *

Values are presented as mean (SE, standard error) or as numbers only. Effects were analyzed using the linear mixed model and calculated as amount per month. * Statistical significance was noted. † The reference value is set to 0 to compare the effect of each factor among the parameters of tibiofibular alignment.

4. Discussion

In this study, we compared the longitudinal changes between the tibiofibular relationship and ankle alignment in HME patients. When the valgus angulation of the tibiotalar angle progressed, we found that the tibial growth was relatively greater than the fibular growth and was accompanied by significantly relative fibular shortening in the proximal and distal portions. This is in line with previous observations that have reported that HME of the lower extremities induces a disproportionate shortening of the fibula with respect to the tibia, leading to progressive valgus deformities of the ankle joint [10,11,15]. Additionally, the sex of the patient was identified to not affect the growth of the tibia or fibula, and synostoses was shown to slow the growth of both the tibia and fibula.

Although the absolute growth rates of the fibula cannot be directly compared, it may be assumed that the relative retardation of fibular growth caused the ankle valgus deformity and fibular shortening [4,13]. For children aged 7–10 years, during physiological growth, the proximal and distal tibiofibular physis differences are usually measured as 1.2 to 1.7 cm and 0.7 to 1.0 cm, respectively. However, in our study, the proximal and distal tibiofibular physis differences were greater than this, and, therefore, we assumed that the fibular growth retardation was apparent [13]. A similar phenomenon may also occur in the forearm, as deformities are caused by relative ulnar shortening [16]. The size of the osteochondromas in the ulna has been inversely correlated with the rate of ulnar shortening, and the location of osteochondromas on the distal ulnar metaphysis has been shown to restrict longitudinal bone growth to under 20% of their expected growth [17]. Using a longitudinal observation study design, we found that the fibula growth slopes were similar regardless of the direction of the change in the ankle alignment. However, for the tibia, significantly greater growth slopes were noted when ankle valgus angulation was present. Therefore, these changes in the tibiofibular relationship of HME patients support the effectiveness of surgical treatment of the tibia with procedures such as medial malleolar screw epiphysiodesis [8,9].

In studies investigating tibial and fibular growth in normal children, progressive distal migration of the distal fibular epiphysis and physis occurred with increasing age [3,13,18,19]. In cases of pediatric ankle fractures or iatrogenic tibiofibular synostosis, it has been reported that the tibiofibular synostosis blocks the physiological distal migration of the fibula, resulting in a proximal fibular migration and deformity with a prominent fibular head and shortened lateral malleolus [5,20–23]. However, our study revealed continuous shortening of both the proximal and distal fibula over time. In addition, the presence of synostosis was found to inhibit the growth of both the tibia and fibula when compared to the growth rates of those without synostoses. Therefore, these findings suggest that the fibular shortening seen in HME patients appears to show different trends with those observed by fractures or iatrogenic synostosis [21].

Takikawa et al. utilized a longitudinal study design with serial ankle radiographs and reported that an ankle valgus deformity could be predicted in male HME patients with prominent shortening of the fibula and exostoses involvement of both the lateral distal tibia and the medial distal fibula [11]. However, they evaluated a total of 62 ankle radiographs of 33 patients (23 males, 10 females) with a mean age of 11.33 years and did not conduct any statistical analysis related to the passage of time. In contrast, we found no significant fixed effects of sex on the growth of the tibia and fibula and no differences between sex or location of the exostosis on the ankle valgus deformity progression. No significant difference was found in age and sex between our groups, and it is possible that the difference in patient distributions compared to the study of Takikawa et al. may have led to the differing results.

A study evaluating the effect of osteochondroma locations on coronal lower limb malalignment reported that the combined lesions of both the proximal and distal tibiofibular joints had the greatest impact on fibular shortening [10]. In our study, the analysis of the combination lesion was not included, and no significance was found in the location of lesions in relation to fibular shortening, regardless of whether valgus progressed or not. Osteochondromas develop and grow while the physis is still open, and with variations in sex, they grow proportionately to the overall growth of the patient [19].

Therefore, bone age may influence the tibiofibular growth and relationship. Our study investigated the alignment changes over time within a cohort of patients all almost under the age of 10, and, hence, there may be differences between our results and those of the cross-sectional study of Ahn et al. on patients with a mean age of 10–14 years [10].

The initial ankle alignment did not significantly affect the future growth of the tibia and fibula in our study. Despite variations with age, minor valgus alignments of 0–8 degrees for the tibiotalar angle are within the standard range during skeletal development [13]. Noonan et al. investigated the natural history of 38 adult patients, mean age of 42 years, with lower extremity and ankle joint HMEs. Their mean tibiotalar angle was 8.6°, half of their cohort (50%) suffered from occasional ankle pain, and 14 of 75 ankles (19%) showed degenerative changes [6]. Although physiologic valgus alignment may be acceptable during the growing period, ankle valgus deformities that persists into adulthood can progress to arthritis and result in deterioration of the ankle joint function. More than 80% of patients with HME are initially diagnosed in the first decade of life. Therefore, careful observation and appropriately timed treatment during the growth period are required to minimize functional impairment and the progression of deformities [4,6].

In a retrospective review study of 113 HME patients, the distribution of osteochondromas was reported to be more likely in the proximal tibia (71%) than the distal tibia, and the case was similar for the proximal and distal fibula (27%) [24]. We also found that the location of osteochondromas occurred more proximally in both the tibia and fibula, a finding which is similar to those previous reported. Therefore, as osteochondromas may arise more proximally, evaluations of full length tibia and fibula radiographs may more reliably predict the progression of ankle valgus than those of ankle radiographs only.

This study has several limitations. First, the duration of follow-up was relatively short, and the follow-up time interval was irregular due to the retrospective study design. A prospective study with extended observations to adulthood and controlled time intervals may provide more reliable information to further our understanding of the progression of angular malalignment of lower extremities during the growth period for HME patients. Second, radiographic evaluation of patients was exclusively performed by teleoroentgenogram, with the center of the beam projection facing the knee. Therefore, there may be an inherent error with this beam projection set-up, and all values should have been compared with simultaneous ankle AP radiographs [25]. Nevertheless, since this study investigated the changes in the tibiofibular relationship over time, depending on the progression of ankle valgus deformity, we focused more on analyzing the overall trends of radiographic changes longitudinally. Last, the accuracy of the observations for the presence of synostosis and evaluation of osteochondroma sizes were insufficient due to the lack of a tomographic examinations. Tomographic scanning has the advantage of accurately demonstrating the condition, location, and synostosis of osteochondroma lesions [20]. However, it is clinically difficult to implement and justify such imaging in the pediatric HME population.

5. Conclusions

In pediatric HME patients, when valgus deformity of the ankle joint progressed, greater relative tibial growth and significantly relative proximal and distal fibular shortening followed. For the fixed effect of characteristic factors on the tibiofibular alignment over time, tibiofibular synostosis lowered the growth of both the tibia and fibula, whilst initial ankle alignment partially affected the tibiofibular relationship. However, demographic and anatomical factors such as age, sex, location and synostosis were not associated with the progression of ankle valgus deformities. Our observations suggest that pediatric HME patients with an imbalanced tibiofibular relationship, due to greater tibial growth, may require appropriate surgical treatment to prevent the progression of ankle valgus deformities, regardless of predisposing factors.

Author Contributions: Conceptualization, J.H.L. and H.W.K.; methodology, J.H.L., H.P. and H.W.K; formal analysis, H.S.L.; investigation, J.H.L. and C.M.R.; data curation, J.H.L. and C.M.R.; writing—original draft preparation, J.H.L.; writing—review and editing, I.R. and H.W.K.; visualization, H.W.K.; supervision, H.W.K.; project administration, H.W.K. All authors have read and agreed to the published version of the manuscript.

Funding: This research received no external funding.

Conflicts of Interest: The authors declare no conflict of interest.

References

1. Shapiro, F.; Simon, S.; Glimcher, M.J. Hereditary multiple exostoses. Anthropometric, roentgenographic, and clinical aspects. *JBJS* **1979**, *61*, 815–824.
2. Snearly, W.N.; Peterson, H.A. Management of ankle deformities in multiple hereditary osteochondromata. *J. Pediatric Orthop.* **1989**, *9*, 427–432. [CrossRef]
3. Solomon, L. Bone growth in diaphysial aclasis. *J. Bone Jt. Surg. Br. Vol.* **1961**, *43-B*, 700–716. [CrossRef]
4. Stieber, J.R.; Dormans, J.P. Manifestations of hereditary multiple exostoses. *JAAOS J. Am. Acad. Orthop. Surg.* **2005**, *13*, 110–120. [CrossRef] [PubMed]
5. Chin, K.R.; Kharrazi, F.D.; Miller, B.S.; Mankin, H.J.; Gebhardt, M.C. Osteochondromas of the distal aspect of the tibia or fibula. Natural history and treatment. *JBJS* **2000**, *82*, 1269–1278. [CrossRef] [PubMed]
6. Noonan, K.J.; Feinberg, J.R.; Levenda, A.; Snead, J.; Wurtz, L.D. Natural history of multiple hereditary osteochondromatosis of the lower extremity and ankle. *J. Pediatric Orthop.* **2002**, *22*, 120–124. [CrossRef]
7. Bozkurt, M.; Dogan, M.; Turanli, S. Osteochondroma leading to proximal tibiofibular synostosis as a cause of persistent ankle pain and lateral knee pain: A case report. *Knee Surg. Sports Traumatol. Arthrosc.* **2004**, *12*, 152–154. [CrossRef]
8. Rupprecht, M.; Spiro, A.S.; Rueger, J.M.; Stucker, R. Temporary screw epiphyseodesis of the distal tibia: A therapeutic option for ankle valgus in patients with hereditary multiple exostosis. *J. Pediatric Orthop.* **2011**, *31*, 89–94. [CrossRef]
9. Rupprecht, M.; Spiro, A.S.; Schlickewei, C.; Breyer, S.; Ridderbusch, K.; Stucker, R. Rebound of ankle valgus deformity in patients with hereditary multiple exostosis. *J. Pediatric Orthop.* **2015**, *35*, 94–99. [CrossRef]
10. Ahn, Y.S.; Woo, S.H.; Kang, S.J.; Jung, S.T. Coronal malalignment of lower legs depending on the locations of the exostoses in patients with multiple hereditary exostoses. *BMC Musculoskelet. Disord.* **2019**, *20*, 564. [CrossRef]
11. Takikawa, K.; Haga, N.; Tanaka, H.; Okada, K. Characteristic factors of ankle valgus with multiple cartilaginous exostoses. *J. Pediatric Orthop.* **2008**, *28*, 761–765. [CrossRef] [PubMed]
12. Sabharwal, S.; Zhao, C. Assessment of lower limb alignment: Supine fluoroscopy compared with a standing full-length radiograph. *JBJS* **2008**, *90*, 43–51. [CrossRef] [PubMed]
13. Beals, R.K.; Skyhar, M. Growth and development of the tibia, fibula, and ankle joint. *Clin. Orthop. Relat. Res.* **1984**, *182*, 289–292. [CrossRef]
14. Lachin, J.M. The role of measurement reliability in clinical trials. *Clin. Trials* **2004**, *1*, 553–566. [CrossRef] [PubMed]
15. Czajka, C.M.; DiCaprio, M.R. What is the Proportion of Patients with Multiple Hereditary Exostoses Who Undergo Malignant Degeneration? *Clin. Orthop. Relat. Res.* **2015**, *473*, 2355–2361. [CrossRef]
16. Akita, S.; Murase, T.; Yonenobu, K.; Shimada, K.; Masada, K.; Yoshikawa, H. Long-term results of surgery for forearm deformities in patients with multiple cartilaginous exostoses. *JBJS* **2007**, *89*, 1993–1999. [CrossRef]
17. Porter, D.E.; Emerton, M.E.; Villanueva-Lopez, F.; Simpson, A.H. Clinical and radiographic analysis of osteochondromas and growth disturbance in hereditary multiple exostoses. *J. Pediatric Orthop.* **2000**, *20*, 246–250. [CrossRef]
18. Karrholm, J.; Hansson, L.I.; Selvik, G. Longitudinal growth rate of the distal tibia and fibula in children. *Clin. Orthop. Relat. Res.* **1984**, *191*, 121–128.
19. Pritchett, J.W. Growth and growth prediction of the fibula. *Clin. Orthop. Relat. Res.* **1997**, *334*, 251–256. [CrossRef]
20. Bessler, W.; Eich, G.; Stuckmann, G.; Zollikofer, C. Kissing osteochondromata leading to synostoses. *Eur. Radiol.* **1997**, *7*, 480–485. [CrossRef]
21. Frick, S.L.; Shoemaker, S.; Mubarak, S.J. Altered fibular growth patterns after tibiofibular synostosis in children. *JBJS* **2001**, *83*, 247–254. [CrossRef] [PubMed]

22. Park, H.W.; Kim, H.W.; Kwak, Y.H.; Roh, J.Y.; Lee, J.J.; Lee, K.S. Ankle valgus deformity secondary to proximal migration of the fibula in tibial lengthening with use of the Ilizarov external fixator. *JBJS* **2011**, *93*, 294–302. [CrossRef] [PubMed]
23. Karrholm, J.; Hansson, L.I.; Selvik, G. Changes in tibiofibular relationships due to growth disturbances after ankle fractures in children. *JBJS* **1984**, *66*, 1198–1210. [CrossRef]
24. Schmale, G.A.; Conrad, E.U., 3rd; Raskind, W.H. The natural history of hereditary multiple exostoses. *JBJS* **1994**, *76*, 986–992. [CrossRef]
25. Jang, W.Y.; Park, M.S.; Yoo, W.J.; Chung, C.Y.; Choi, I.H.; Cho, T.J. Beam Projection Effect in the Radiographic Evaluation of Ankle Valgus Deformity Associated with Fibular Shortening. *J. Pediatric Orthop.* **2016**, *36*, e101–e105. [CrossRef]

Article

Relationship between Cortical Bone Thickness and Cancellous Bone Density at Dental Implant Sites in the Jawbone

Shiuan-Hui Wang [1], Yen-Wen Shen [2,3], Lih-Jyh Fuh [2,3], Shin-Lei Peng [4], Ming-Tzu Tsai [5], Heng-Li Huang [2,6] and Jui-Ting Hsu [2,6,*]

1 Master Program for Biomedical Engineering, China Medical University, Taichung 404, Taiwan; u104020415@cmu.edu.tw
2 School of Dentistry, China Medical University, Taichung 404, Taiwan; a2312830@ms28.hinet.net (Y.-W.S.); ljfuh@mail.cmu.edu.tw (L.-J.F.); hlhuang@mail.cmu.edu.tw (H.-L.H.)
3 Department of Dentistry, China Medical University and Hospital, Taichung 404, Taiwan
4 Department of Biomedical Imaging and Radiological Science, China Medical University, Taichung 404, Taiwan; speng@mail.cmu.edu.tw
5 Department of Biomedical Engineering, Hungkuang University, Taichung 433, Taiwan; anniemtt@hk.edu.tw
6 Department of Bioinformatics and Medical Engineering, Asia University, Taichung 413, Taiwan
* Correspondence: jthsu@mail.cmu.edu.tw

Received: 19 August 2020; Accepted: 15 September 2020; Published: 17 September 2020

Abstract: Dental implant surgery is a common treatment for missing teeth. Its survival rate is considerably affected by host bone quality and quantity, which is often assessed prior to surgery through dental cone-beam computed tomography (CBCT). Dental CBCT was used in this study to evaluate dental implant sites for (1) differences in and (2) correlations between cancellous bone density and cortical bone thickness among four regions of the jawbone. In total, 315 dental implant sites (39 in the anterior mandible, 42 in the anterior maxilla, 107 in the posterior mandible, and 127 in the posterior maxilla) were identified in dental CBCT images from 128 patients. All CBCT images were loaded into Mimics 15.0 to measure cancellous bone density (unit: grayscale value (GV) and cortical bone thickness (unit: mm)). Differences among the four regions of the jawbone were evaluated using one-way analysis of variance and Scheffe's posttest. Pearson coefficients for correlations between cancellous bone density and cortical bone thickness were also calculated for the four jawbone regions. The results revealed that the mean cancellous bone density was highest in the anterior mandible (722 ± 227 GV), followed by the anterior maxilla (542 ± 208 GV), posterior mandible (535 ± 206 GV), and posterior maxilla (388 ± 206 GV). Cortical bone thickness was highest in the posterior mandible (1.15 ± 0.42 mm), followed by the anterior mandible (1.01 ± 0.32 mm), anterior maxilla (0.89 ± 0.26 mm), and posterior maxilla (0.72 ± 0.19 mm). In the whole jawbone, a weak correlation (r = 0.133, p = 0.041) was detected between cancellous bone density and cortical bone thickness. Furthermore, except for the anterior maxilla (r = 0.306, p = 0.048), no correlation between the two bone parameters was observed (all $p > 0.05$). Cancellous bone density and cortical bone thickness varies by implant site in the four regions of the jawbone. The cortical and cancellous bone of a jawbone dental implant site should be evaluated individually before surgery.

Keywords: jawbone; cancellous bone density; cortical bone thickness; dental cone-beam computed tomography; dental implant site

---

1. Introduction

Dental implant surgery is a common treatment for missing teeth [1,2]. Its survival rate is closely related to osseointegration ability in patients. Specifically, dental implants with higher initial stability have more favorable osseointegration and are associated with a higher survival rate [3–7]; accordingly, examining the bone condition of a dental implant site in the jawbone is crucial. Generally, jawbone condition is assessed by measuring cancellous bone density and cortical bone thickness.

Numerous studies in dentistry have assessed jawbone condition by using computed tomography (CT) and dental cone-beam CT (CBCT), and have evaluated cancellous bone density prior to dental implant surgery on the basis of bone radiographic density. For CT and dental CBCT images, bone density at a dental implant site can be expressed as Hounsfield units (HU) and grayscale values (GVs) [8–12], respectively. Bone density is generally higher in the mandible and anterior region than it is in the maxilla and posterior regions [13,14]. Studies on bone density at different dental implant sites using CT and CBCT for assessment have reported the following ranking (in descending order) of bone density in different regions: anterior mandible, anterior maxilla, posterior mandible, and posterior mandible [9–12,15].

In addition to cancellous bone density, cortical bone thickness in the jawbone affects the initial stability of dental implants [4,16–20]. Miyamoto et al. [7] employed resonance frequency analyses to measure the initial stability of 225 dental implants; their results demonstrated that implant sites with thicker cortical bone had higher initial stability. Song et al. [21] reported that cortical bone thickness measured using dental CBCT at implant sites was highly correlated with the initial stability of dental implants. Hsu et al. [4] conducted artificial bone experiments and demonstrated that cortical bone thickness significantly influenced the initial stability of dental implants. Roze et al. [22] examined jawbone structures by using micro-CT and compared the results with the initial stability of dental implants. They discovered that dental implant stability depended greatly on cortical bone thickness. In brief, studies on cancellous bone density, cortical bone thickness, and the success rate of dental implant surgery have indicated that jawbone condition is highly correlated with the survival rate of dental implant surgery. In 2017, Ko et al. [23] analyzed the cortical bone thickness of 661 dental implant sites in 173 patients by using dental CBCT and reported the following rankings (in descending order) for cortical bone thickness in different regions: posterior mandible, anterior mandible, anterior maxilla, and posterior maxilla. In the same year, Gupta et al. [24] measured cortical bone thickness at dental implant sites by using dental CBCT and obtained a similar result.

Research has indicated that cancellous bone density and cortical bone thickness at dental implant sites differ among distinct jawbone regions; however, the posterior maxilla has the lowest cancellous bone density and cortical bone thickness. Researchers have examined cancellous bone density [9–12,15] and cortical bone thickness [23,24] by using CT and dental CBCT; however, few have analyzed the correlation between the two parameters. Therefore, dental CBCT was employed in this study to identify (1) differences in and (2) correlations between cancellous bone density and cortical bone thickness at dental implant sites in different jawbone regions.

2. Materials and Methods

2.1. Dental CBCT Examinations of Patients and Implant Sites

This study was conducted after receiving ethical approval from the Institutional Review Board of China Medical University Hospital (No. CMUH 108-REC2-083), approval date: 3 July 2019. Dental CBCT images were obtained from 128 patients (66 male patients, age: 54.14 ± 14.40 years (mean ± standard deviation); 62 female patients, age: 52.13 ± 13.73 years) who had received dental implants between August 2018 and March 2020. Dental CBCT was performed using a Promax 3D Max (Planmeca, Helsinki, Finland) with the following technical parameters: 96 kV, 12.5 mA, and a voxel resolution of 150 or 200 μm. In total, 315 dental implant sites (39 in the anterior mandible, 42 in the anterior maxilla, 107 in the posterior mandible, and 127 in the posterior maxilla) were identified in the dental CBCT images.

2.2. Measurement of Cancellous Bone Density at Dental Implant Sites

The CBCT images were loaded into Mimics (Materialise, Leuven, Belgium). Cylinders (3.5–4 mm in diameter and 8–12 mm in length) were virtually defined according to the dental implant site. The size of each cylinder corresponded to the subsequently inserted dental implant. Cancellous bone density was quantified as the mean GV of the cylinder (Figure 1).

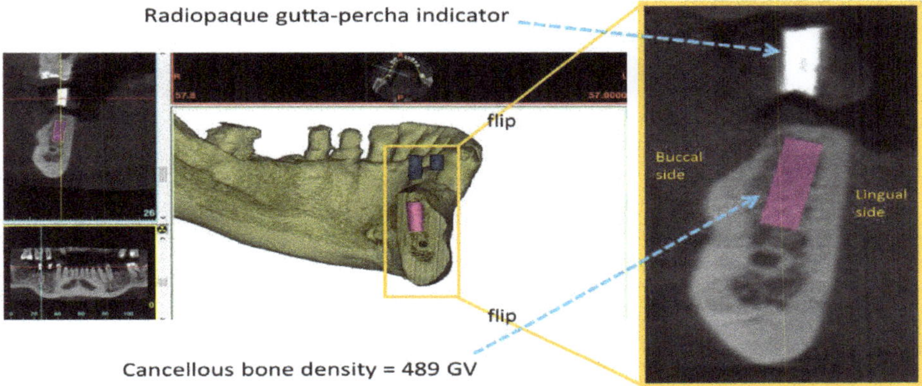

Figure 1. Measurement of cancellous bone density, quantified as the mean grayscale values (GVs) of the cylinder, at a dental implant site.

2.3. Measurement of Cortical Bone Thickness at Dental Implant Sites

In Mimics, continual buccolingual (cross-sectional) images of the maxilla or mandibular bone were captured using the "online reslice" function. To accurately identify the insertion position for a dental implant, diagnostic surgical guide stents with radiopaque gutta-percha indicators were applied for each patient before dental CBCT. Cortical bone thickness was measured in the buccolingual image of the dental implant site, which was made visible by radiopaque gutta-percha indicators (Figure 2).

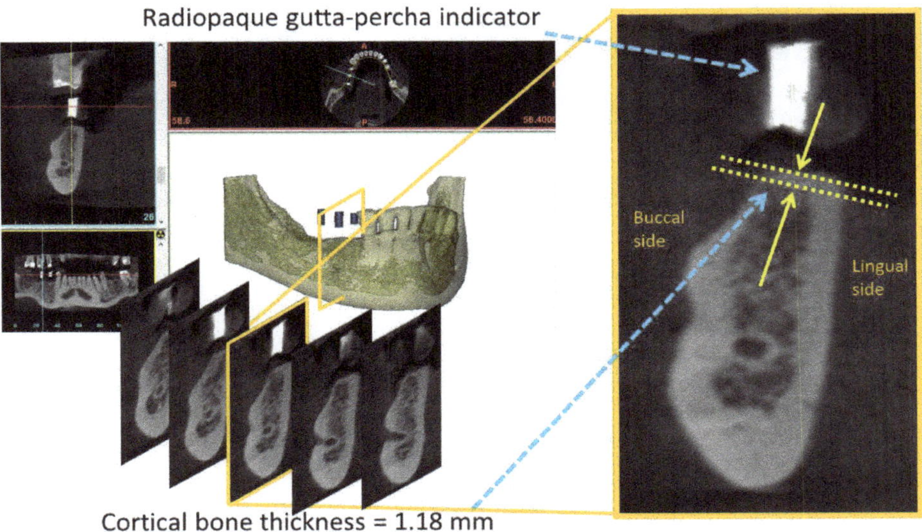

Figure 2. Measurement of cortical bone thickness, distance between the two yellow arrows, at a dental implant site.

2.4. Statistical Analysis

All statistical analyses were conducted using SPSS Version 19 (IBM Corporation, Armonk, NY, USA) and the significance level was set to $p < 0.05$. The following two statistical methods were used to assess the objectives investigated in this study:

(1) Cancellous bone density or cortical bone thickness differences at dental implant sites in four regions of the jawbone (i.e., the anterior maxilla, posterior maxilla, anterior mandible, and posterior mandible) were analyzed using a one-way analysis of variance (ANOVA) and Scheffe's post hoc test for multiple comparisons.

(2) The relationship between cancellous bone density and cortical bone thickness at dental implant sites was analyzed using the Pearson correlation coefficients (r) for the four regions of the jawbone.

3. Results

3.1. Cancellous Bone Density and Cortical Bone Thickness at Dental Implant Sites

For the 315 dental implant sites, cancellous bone density was highest in the anterior mandible (722 ± 227GV) followed by the anterior maxilla (542 ± 208 GV), posterior mandible (535 ± 206 GV), and posterior maxilla (388 ± 206 GV) (Figure 3.). Except for the difference between the posterior mandible and anterior maxilla ($p = 0.098$), the ANOVA and post hoc Scheffe's test indicated that the cancellous bone density was significantly different among the four regions of the jawbone.

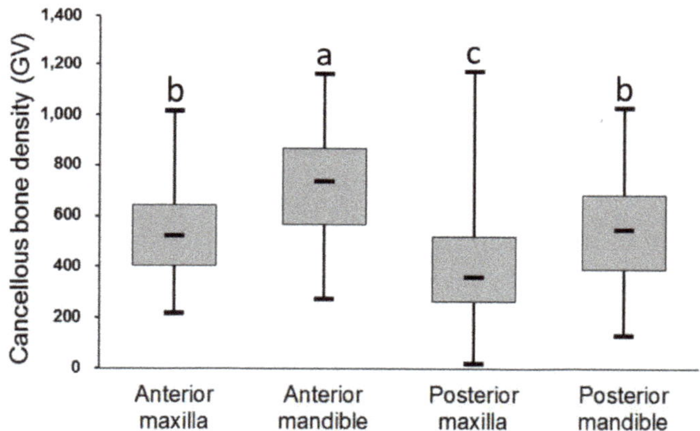

Figure 3. Cancellous bone density at dental implant sites in different regions of the jawbone. Post hoc pairwise comparisons were conducted using Scheffe's test; use of the same letter indicates no significant difference at the 0.05 level.

For the 315 dental implant sites, cortical bone thickness was highest in the posterior mandible (1.15 ± 0.42 mm) followed by the anterior mandible (1.01 ± 0.32 mm), anterior maxilla (0.89 ± 0.26 mm), and posterior maxilla (0.72 ± 0.19 mm) (Figure 4). Except for the differences between the anterior maxilla and anterior mandible ($p = 0.445$) and between the anterior mandible and posterior mandible ($p = 0.110$), the ANOVA and post hoc Scheffe's test indicated that cortical bone thickness differed significantly among the four regions of the jawbone.

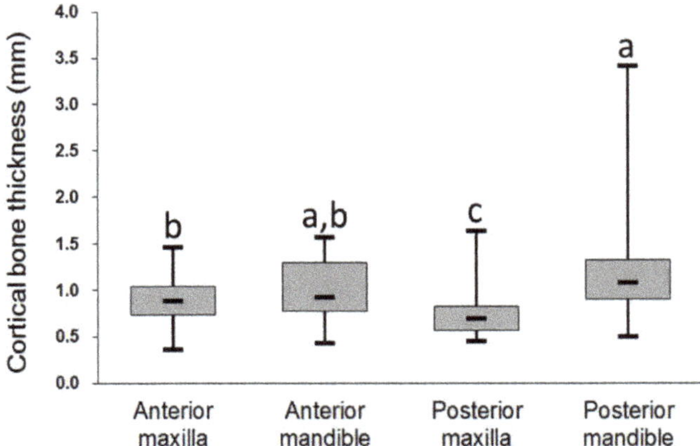

Figure 4. Cortical bone thickness at dental implant sites in different regions of the jawbone. Post hoc pairwise comparisons were conducted using Scheffe's test; use of the same letter indicates no significant difference at the 0.05 level.

3.2. *Relationship between Cancellous Bone Density and Cortical Bone Thickness at Dental Implant Sites*

For the entire jawbone, a weak correlation between cancellous bone density and cortical bone thickness was detected (r = 0.133, p = 0.041, Table 1) for the 315 dental implant sites. In addition, except for the maxilla region (r = 0.168, p = 0.029), posterior region (r = 0.178, p = 0.006), and anterior maxilla (r = 0.306, p = 0.048), which all had weak correlations, none of the other regions exhibited correlations between cancellous bone density and cortical bone thickness (Table 1).

Table 1. Relationship between cancellous bone density and cortical bone thickness at dental implant sites.

Region of the Jawbone	Numbered	Cancellous Bone Density (Mean ± SD)	Cortical Bone Thickness (Mean ± SD)	Pearson Correlation Coefficient	
				r	p
All region	315	500 ± 235	0.92 ± 0.36	0.133	0.041
Mandible	146	585 ± 227	1.11 ± 0.40	−0.125	0.133
Maxilla	169	426 ± 217	0.76 ± 0.23	0.168	0.029
Anterior	81	629 ± 235	0.95 ± 0.30	0.049	0.662
Posterior	234	456 ± 219	0.92 ± 0.38	0.178	0.006
Anterior maxilla	42	542 ± 208	0.89 ± 0.26	0.306	0.048
Anterior mandible	39	722 ± 227	1.01 ± 0.32	−0.296	0.068
Posterior maxilla	127	388 ± 206	0.72 ± 0.19	−0.036	0.686
Posterior mandible	107	535 ± 206	1.15 ± 0.42	−0.006	0.952
Unit		GV (Grayscale value)	mm		

4. Discussion

Dental implant surgery has gradually become the prevailing treatment for missing teeth and the assessment of jawbone condition is an essential procedure prior to such surgery. The relevant literature has indicated that a favorable bone condition results in stronger osseointegration and higher dental implant survival rates [25,26]. Currently, cancellous bone density and cortical bone thickness are the most common indicators used to determine bone condition [3,5–7,21,22,27]. Researchers have measured the cancellous bone density of jawbones by using CT and dental CBCT [8–12]; however, few have evaluated cortical bone thickness at dental implant sites [23,24], and nearly none have explored the correlation between cortical bone thickness and cancellous bone density at dental implant sites in jawbones. This study was the first to employ dental CBCT to assess patients' cancellous bone density and cortical bone thickness at dental implant sites and further examine the correlation between the two parameters. Consistent with the relevant literature, this study verified that cancellous bone density and cortical bone thickness differed among distinct jawbone regions. Furthermore, the results indicate that cortical bone thickness and cancellous bone density in the entire jawbone had a weak correlation; of the four jawbone regions, the two indicators exhibited a weak correlation in the anterior maxilla and nonsignificant correlations in the remaining three regions.

Of studies that have examined the relationship between jawbone condition and the success rate of dental implants, Jaffin et al. [5] demonstrated that in the Lekholm and Zarb bone classification [28], the failure rate of dental implants in Type-IV host bones reached 35%, but that in Type-I–III host bones with higher bone quality was only 3%. Jemt et al. [6] also demonstrated that the failure rate of dental implants in high-quality bones was 7.9%, whereas that in low-quality bones was 28.8%. Moreover, clinical literature has disclosed that implant survival rates are higher when implants are embedded in sites with high bone quality, which provide more initial stability for dental implants, according to clinical research [7] and artificial bone experiments [19,29,30]. Higher initial stability contributes to more favorable osseointegration and a higher implant survival rate [3,5,6,31,32]. Accordingly, bone quality and quantity at implant sites can convey crucial information required prior to dental implant surgery.

Dental CBCT is currently a standard procedure in routine evaluation prior to dental implant surgery. It possesses advantages, such as a low radiation dose and high spatial resolution, and does not create distortion in three-dimensional images or induce image overlapping [33]. Although some researchers have argued that the use of GVs to present dental CBCT results regarding cancellous bone density is imprecise [34,35], other researchers have implied that such a technique is applicable for bone density evaluation because of the maturity of relevant technology [36–40]. Naitoh et al. [39] and Nomura et al. [41] compared CT and dental CBCT in bone density assessment, and both studies concluded that the HU and GV obtained in CT and dental CBCT, respectively, had a strong positive correlation. Parsa et al. [40] compared dental CBCT with CT and micro-CT for measuring bone tissue, and the results indicated that the GV obtained in dental CBCT was highly correlated with the HU and bone volume fraction obtained in CT and micro-CT, respectively. In 2017, Liu et al., [37] indicated that dental CBCT images effectively indicate cancellous bone density and are a suitable instrument for assessment prior to dental implant surgery. Hao et al., [11], David et al. [9], and Naitoh et al. [39] also employed dental CBCT to measure bone density at dental implant sites.

Numerous researchers have employed CT and dental CBCT to measure cancellous bone density at dental implant sites in jawbones. Of the researchers who have used CT to evaluate the density in the four jawbone regions, Turkyilmaz et al. [12] and de Olivira et al. [15] indicated that bone density differs significantly among regions and is ranked as follows in descending order: anterior mandible, anterior maxilla, posterior mandible, and posterior maxilla. Moreover, Shapurian et al. [14] and Fuh et al. [10] have indicated that the differences among the four jawbone regions are not always statistically significant; nevertheless, the bone density of the mandible and anterior region is generally higher than that of the maxilla and posterior region. Researchers have used dental CBCT as an assessment instrument for bone density at implant sites and reached similar conclusions. According to Hao et al. [11], bone density descended in the following order: anterior mandible (680 ± 142 GV) > anterior maxilla (460 ± 136 GV), posterior mandible (394 ± 128 GV) > posterior maxilla (230 ± 144 GV), which corresponded to the following result of David et al. [9]: anterior mandible (female: 514 ± 243 GV, male: 521 ± 247 GV) > anterior maxilla (female: 354 ± 206 GV, male: 473 ± 208 GV), posterior mandible (female: 234 ± 212 GV, male: 389 ± 220 GV) > posterior maxilla (female: 193 ± 176 GV, male: 250 ± 193 GV). The present study obtained a similar result; cancellous bone density in different regions was ranked as follows in descending order: anterior mandible (722 ± 227 GV), anterior maxilla (542 ± 208 GV), posterior mandible (535 ± 206 GV), and posterior maxilla (388 ± 206 GV). The different dental CBCT models and scanning parameters (e.g., voltage, current, time, and resolution) and the inclusion of patients of a different ethnicity might have contributed to differences in cancellous bone densities between this study and preceding studies [10]. In sum, generally, the results of this and preceding studies demonstrate that cancellous bone density at dental implant sites in jawbones descended in the following order: anterior mandible, anterior maxilla, posterior mandible, and posterior maxilla.

Previous studies evaluating cortical bone thickness by using dental CBCT have mostly focused on the thickness on the buccal and palatal sides in orthodontic patients [42–44]; few have emphasized crestal cortical bone thickness. In 2013, Gerlach et al. [45] employed dental CBCT to evaluate crestal cortical bone thickness at six mandibular dental implant sites in a patient with missing teeth and obtained a thickness of 2.00 ± 0.15 mm. However, crestal cortical bone thickness might be overestimated because of the partial volume effects caused by the 400 μm resolution in dental CBCT. In 2017, Ko et al. [23] adopted dental CBCT with a resolution of 155 μm to evaluate crestal cortical bone thickness at dental implant sites in the four jawbone regions and revealed that thickness was ranked as follows in descending order: posterior mandible (1.22 ± 0.52 mm) > anterior mandible (1.06 ± 0.32 mm) > anterior maxilla (0.83 ± 0.31 mm) > posterior maxilla (0.72 ± 0.29 mm). In the same year, Gupta et al. [24] implemented similar experiments and obtained a similar order of posterior mandible (1.18 ± 0.48 mm) > anterior mandible (1.08 ± 0.30 mm) > anterior maxilla (0.82 ± 0.32 mm) > posterior maxilla (0.76 ± 0.29 mm). The study of Gupta et al. [24] was based in India, and patient ethnicity was mainly Caucasian and Australian. The experimental results of Ko et al., [23] and the present study were similar because patients in both studies had Mongolian ethnicity. In brief, the present and preceding studies have all indicated that crestal cortical bone thickness at dental implant sites is greatest in the posterior mandible, followed by the anterior mandible, anterior maxilla, and posterior maxilla.

This study analyzed the correlation between cancellous bone density and cortical bone thickness at dental implant sites by using the Pearson correlation coefficient. The statistical results for the different jawbone regions demonstrated that the two parameters had no correlation in the posterior maxilla, anterior mandible, or posterior mandible ($p > 0.05$) but had a weak correlation in the anterior maxilla ($r = 0.306$, $p = 0.048$), with p approaching 0.05. Notably, when the jawbones were divided into only anterior and posterior regions, a significant but weak correlation was observed in the posterior region ($r = 0.178$, $p = 0.006$); when the jawbones were classified into only maxilla and mandible regions, a significant but weak correlation was also observed in the maxilla region ($r = 0.168$, $p = 0.029$). Furthermore, the results for the entire jawbone indicate that cancellous bone density and cortical bone thickness had a weak correlation ($r = 0.133$, $p = 0.041$).

Compared with parameters of cancellous bone density, crestal cortical bone thickness parameters in the anterior mandible were not optimal in the four jawbone regions. Crestal cortical bone thickness may have been lower in anterior mandible than in the posterior mandible for the following reasons: (1) The mandible is categorized as a class III lever; to balance mandible and total blood weight under functional demands, the lighter anterior mandible exerts less energy to achieve mandibular movements than does the heavier anterior mandible [46,47]. (2) Anterior teeth tend to withstand more frequent oblique force than do posterior teeth; specifically, because the anterior mandible frequently tolerates horizontal force, it requires high-density cancellous bone to resist the horizontal force transmitted from the roots to the surrounding bones. (3) Masseter muscles are attached to the buccal side of mandible, but jawbone growth increases crestal cortical bone thickness. Although Cassetta et al. [48] measured crestal cortical bone thickness on the lingual and buccal sides, they reported that the cortical bones in the posterior region were thicker than those in the anterior region. In addition, regardless of crestal cortical bone thickness or cancellous bone density, the posterior maxilla had the lowest bone quality and quantity because it is located near the maxillary sinus. Accordingly, tooth loss often results in pneumatization, deteriorates the bone in the posterior maxilla, reduces cancellous bone density, and reduces cortical bone thickness. These reasons correspond with the results of previous studies [23,24].

The experimental results of this study revealed that crestal cortical bone thickness and cancellous bone density had different rankings at dental implant sites in the four jawbone regions. The posterior maxilla had the lowest cancellous bone density and the thinnest crestal cortical bone. However, no correlation was observed between the crestal cortical bone thickness and cancellous bone density in this region ($p > 0.05$). Figure 5a presents scatter plots of cortical bone thickness and cancellous bone density of the anterior maxilla and posterior maxilla, respectively, of all patients. The two indicators were not correlated in the posterior maxilla ($p > 0.05$); however, because cancellous bone density and crestal cortical bone thickness in the posterior maxilla were much lower than those in the anterior maxilla, the two indicators were weakly correlated in the maxilla region ($r = 0.168$, $p = 0.029$). Similarly, the two indicators were not correlated in the posterior maxilla or posterior mandible alone ($p > 0.05$), but the overall values in the posterior maxilla were lower than those in the posterior mandible. Therefore, the two indicators demonstrated a weak correlation in the posterior region ($r = 0.178$, $p = 0.006$, Figure 5b).

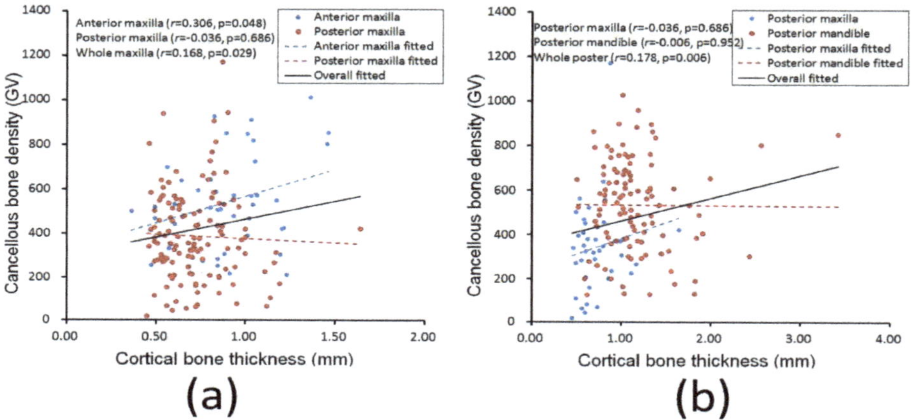

Figure 5. Relationships between cortical bone thickness and cancellous bone density at dental implant sites in the (**a**) anterior maxilla (blue dots and dotted line), posterior maxilla (red dots and dotted line), and the entire maxilla region (black line) and in the (**b**) posterior maxilla (blue dots and dotted line), posterior mandible (red dots and dotted line), and entire posterior region (black line).

Numerous studies have verified that cancellous bone density and crestal cortical bone thickness at dental implant sites greatly influence the success rate of dental implant surgery [3,5–7,21,22,27] because dental implants with superior bone conditions have greater initial stability [3,5–7]. However, cortical and cancellous bone have different roles in maintaining the stability of dental implants [22,27]. Cortical bone is generally related to the initial stability of dental implants, whereas cancellous bone, which is composed of trabecular bone and is filled with more blood, is more related to osseointegration in subsequent stages. The most common clinical approach for assessing bone quality and quantity in the jaw is the use of the Lekholm and Zarb bone classification [28]. Jawbones are categorized into four types from most to least preferable, and this classification method was established on the basis of the premise that thicker cortical bone results in denser cancellous bone, whereas thinner cortical bone contributes to looser cancellous bone. Nevertheless, this study revealed that crestal cortical bone thickness and cancellous bone density differed inconsistently at the dental implant sites in the four jawbone regions; furthermore, cancellous bone density and cortical bone thickness were weakly correlated and even exhibited no correlation in some regions. This suggests that thicker cortical bone is not necessarily associated with denser cancellous bone. Accordingly, for patients requiring dental implants, dentists can assess the cortical and cancellous bone conditions at implant sites by using dental CBCT to develop comprehensive preoperative plans for dental implants.

This study had the following limitations: (1) All patients recruited in this study were Asian; whether these results are generalizable to other ethnic groups requires further examination. (2) Patients in this study were not grouped by sex or age because the sample size was insufficient. (3) Cancellous bone density was measured as radiographic density (in GVs) in this study. However, several researchers have indicated that dental CBCT is not as accurate as clinical CT is for measuring bone density. In the future, a calibration phantom should be employed alongside CBCT for calculations of bone mineral density at a dental implant site. (4) The patients selected in this study were carefully evaluated by the dentist and judged to be suitable candidates for dental implant surgery. This study did not include the effects of specific patients' conditions (e.g., how long patients used removable prostheses, time from tooth extraction, and patients' physical conditions). (5) This study examined only patients' dental CBCT images prior to their dental implant surgery and did not perform follow-ups to determine survival rates after surgery. In subsequent research, we plan to implement more comprehensive assessments of the influence of jawbone condition on dental implant survival rates.

5. Conclusions

The following conclusions regarding host bone condition at the dental implant site in the four jawbone regions were drawn on the basis of the experimental setup and limitations:

(1) Cancellous bone density was highest in the anterior mandible (722 ± 227 GV), followed by the anterior maxilla (542 ± 208 GV), posterior mandible (535 ± 206 GV), and posterior maxilla (388 ± 206 GV).

(2) The cortical bone was thickest in the posterior mandible (1.15 ± 0.42 mm), followed by the anterior mandible (1.01 ± 0.32 mm), anterior maxilla (0.89 ± 0.26 mm), and posterior maxilla (0.72 ± 0.19 mm).

(3) A weak correlation ($r = 0.133$, $p = 0.041$) was observed between cancellous bone density and cortical bone thickness in the entire jawbone. Furthermore, except for the anterior maxilla ($r = 0.306$, $p = 0.048$), no correlation between the two bone parameters was detected.

Author Contributions: Conceptualization, S.-H.W. and J.-T.H.; methodology, S.-H.W., Y.-W.S., S.-L.P., and J.-T.H.; writing—original draft preparation, S.-H.W., Y.-W.S., M.-T.T., H.-L.H., and J.-T.H.; writing—review and editing, S.-H.W., L.-J.F., M.-T.T., and J.-T.H.; funding acquisition, J.-T.H. All authors have read and agreed to the published version of the manuscript.

Funding: This study was funded by the China Medical University (CMU108-MF-86) and Ministry of Science and Technology, Taiwan (MOST 109-2221-E-039-005).

Conflicts of Interest: The authors declare no conflict of interest.

References

1. Fueki, K.; Yoshida, E.; Igarashi, Y. A systematic review of prosthetic restoration in patients with shortened dental arches. *Jpn. Dent. Sci. Rev.* **2011**, *47*, 167–174. [CrossRef]
2. Hong, D.G.K.; Oh, J.-H. Recent advances in dental implants. *Maxillofac. Plast. Reconstr. Surg.* **2017**, *39*, 33. [CrossRef] [PubMed]
3. Esposito, M.; Hirsch, J.M.; Lekholm, U.; Thomsen, P. Biological factors contributing to failures of osseointegrated oral implants, (II). Etiopathogenesis. *Eur. J. Oral Sci.* **1998**, *106*, 721–764. [CrossRef] [PubMed]
4. Hsu, J.T.; Fuh, L.J.; Tu, M.G.; Li, Y.F.; Chen, K.T.; Huang, H.L. The effects of cortical bone thickness and trabecular bone strength on noninvasive measures of the implant primary stability using synthetic bone models. *Clin. Implant Dent. Relat. Res.* **2013**, *15*, 251–261. [CrossRef] [PubMed]
5. Jaffin, R.A.; Berman, C.L. The excessive loss of Branemark fixtures in type IV bone: A 5-year analysis. *J. Periodontol.* **1991**, *62*, 2–4. [CrossRef]
6. Jemt, T.; Lekholm, U. Implant treatment in edentulous maxillae: A 5-year follow-up report on patients with different degrees of jaw resorption. *Int. J. Oral Maxillofac. Implants* **1995**, *10*, 303–311.
7. Miyamoto, I.; Tsuboi, Y.; Wada, E.; Suwa, H.; Iizuka, T. Influence of cortical bone thickness and implant length on implant stability at the time of surgery—Clinical, prospective, biomechanical, and imaging study. *Bone* **2005**, *37*, 776–780. [CrossRef]

8. Chun, Y.; Lim, W. Bone density at interradicular sites: Implications for orthodontic mini-implant placement. *Orthod. Craniofacial Res.* **2009**, *12*, 25–32. [CrossRef]
9. David, O.; Leretter, M.; Neagu, A. The quality of trabecular bone assessed using cone-beam computed tomography. *Rom. J. Biophys.* **2014**, *24*, 227–241.
10. Fuh, L.J.; Huang, H.L.; Chen, C.S.; Fu, K.L.; Shen, Y.W.; Tu, M.G.; Shen, W.C.; Hsu, J.T. Variations in bone density at dental implant sites in different regions of the jawbone. *J. Oral Rehabil.* **2010**, *37*, 346–351. [CrossRef]
11. Hao, Y.; Zhao, W.; Wang, Y.; Yu, J.; Zou, D. Assessments of jaw bone density at implant sites using 3D cone-beam computed tomography. *Eur. Rev. Med. Pharmacol. Sci.* **2014**, *1*, D1.
12. Turkyilmaz, I.; McGlumphy, E.A. Influence of bone density on implant stability parameters and implant success: A retrospective clinical study. *BMC Oral Health* **2008**, *8*, 32. [CrossRef] [PubMed]
13. Norton, M.R.; Gamble, C. Bone classification: An objective scale of bone density using the computerized tomography scan. *Clin. Oral Implant. Res.* **2001**, *12*, 79–84. [CrossRef] [PubMed]
14. Shapurian, T.; Damoulis, P.D.; Reiser, G.M.; Griffin, T.J.; Rand, W.M. Quantitative evaluation of bone density using the Hounsfield index. *Int. J. Oral Maxillofac. Implants* **2006**, *21*, 290–297.
15. de Oliveira, R.C.G.; Leles, C.R.; Normanha, L.M.; Lindh, C.; Ribeiro-Rotta, R.F. Assessments of trabecular bone density at implant sites on CT images. *Oral Surg. Oral Med. Oral Pathol. Oral Radiol. Endodontol.* **2008**, *105*, 231–238. [CrossRef]
16. Attard, N.J.; Zarb, G.A. Immediate and early implant loading protocols: A literature review of clinical studies. *J. Prosthet. Dent.* **2005**, *94*, 242–258. [CrossRef]
17. Del Fabbro, M.; Rosano, G.; Taschieri, S. Implant survival rates after maxillary sinus augmentation. *Eur. J. Oral Sci.* **2008**, *116*, 497–506. [CrossRef]
18. Gapski, R.; Wang, H.L.; Mascarenhas, P.; Lang, N.P. Critical review of immediate implant loading. *Clin. Oral Implants Res.* **2003**, *14*, 515–527. [CrossRef]
19. Huang, H.-L.; Tu, M.-G.; Fuh, L.-J.; Chen, Y.-C.; Wu, C.-L.; Chen, S.-I.; Hsu, J.-T. Effects of elasticity and structure of trabecular bone on the primary stability of dental implants. *J. Med. Biol. Eng.* **2010**, *30*, 85–89. [CrossRef]
20. Javed, F.; Romanos, G.E.J. The role of primary stability for successful immediate loading of dental implants. A literature review. *J. Dent.* **2010**, *38*, 612–620. [CrossRef]
21. Song, Y.-D.; Jun, S.-H.; Kwon, J.-J.; Implants, M. Correlation between bone quality evaluated by cone-beam computerized tomography and implant primary stability. *Int. J. Oral Maxillofac. Implants* **2009**, *24*, 59–64. [PubMed]
22. Rozé, J.; Babu, S.; Saffarzadeh, A.; Gayet-Delacroix, M.; Hoornaert, A.; Layrolle, P. Correlating implant stability to bone structure. *Clin. Oral Implants Res.* **2009**, *20*, 1140–1145. [CrossRef] [PubMed]
23. Ko, Y.C.; Huang, H.L.; Shen, Y.W.; Cai, J.Y.; Fuh, L.J.; Hsu, J.T. Variations in crestal cortical bone thickness at dental implant sites in different regions of the jawbone. *Clin. Implant Dent. Relat. Res.* **2017**, *19*, 440–446. [CrossRef] [PubMed]
24. Gupta, A.; Rathee, S.; Agarwal, J.; Pachar, R.B. Measurement of Crestal Cortical Bone Thickness at Implant Site: A Cone Beam Computed Tomography Study. *J. Contemp. Dent. Pract.* **2017**, *18*, 785–789. [CrossRef]
25. Brånemark, P.-I.; Breine, U.; Adell, R.; Hansson, B.; Lindström, J.; Ohlsson, Å. Intra-osseous anchorage of dental prostheses: I. Experimental studies. *Scand. J. Plast. Reconstr. Surg.* **1969**, *3*, 81–100. [CrossRef]
26. Branemark, P.-I. Osseointegrated implants in the treatment of the edentulous jaw. Experience from a 10-year period. *Scand. J. Plast. Reconstr. Surg. Hand Surg. Suppl.* **1977**, *16*, 1–132.
27. Isoda, K.; Ayukawa, Y.; Tsukiyama, Y.; Sogo, M.; Matsushita, Y.; Koyano, K. Relationship between the bone density estimated by cone-beam computed tomography and the primary stability of dental implants. *Clin. Oral Implants Res.* **2012**, *23*, 832–836. [CrossRef]
28. Lekholm, U.; Zarb, G.A. Patient selection and preparation. In *Tissue-Integrated Prostheses: Osseointegration in Clinical Dentistry*; Brånemark, P.I., Zarb, G.A., Albrektsson, T., Eds.; Quintessence: Chicago, IL, USA, 1985; pp. 199–209.
29. Hsu, J.-T.; Huang, H.-L.; Tsai, M.-T.; Wu, A.Y.-J.; Tu, M.-G.; Fuh, L.-J. Effects of the 3D bone-to-implant contact and bone stiffness on the initial stability of a dental implant: Micro-CT and resonance frequency analyses. *Int. J. Oral Maxillofac. Surg.* **2013**, *42*, 276–280. [CrossRef]
30. Huang, H.L.; Chang, Y.Y.; Lin, D.J.; Li, Y.F.; Chen, K.T.; Hsu, J.T. Initial stability and bone strain evaluation of the immediately loaded dental implant: An in vitro model study. *Clin. Oral Implants Res.* **2011**, *22*, 691–698. [CrossRef]

31. Adell, R.; Lekholm, U.; Rockler, B.; Brånemark, P.-I. A 15-year study of osseointegrated implants in the treatment of the edentulous jaw. *Int. J. Oral Surg.* **1981**, *10*, 387–416. [CrossRef]
32. Schnitman, P.; Rubenstein, J.E.; Whorle, P.; DaSilva, J.; Koch, G.G. Implants for partial edentulism. *J. Dent. Educ.* **1988**, *52*, 725–736. [CrossRef] [PubMed]
33. Frederiksen, N.L. Diagnostic imaging in dental implantology. *Oral Surg. Oral Med. Oral Pathol. Oral Radiol. Endodontol.* **1995**, *80*, 540–554. [CrossRef]
34. Silva, I.M.d.C.C.; Freitas, D.Q.d.; Ambrosano, G.M.B.; Bóscolo, F.N.; Almeida, S.M. Bone density: Comparative evaluation of Hounsfield units in multislice and cone-beam computed tomography. *Braz. Oral Res.* **2012**, *26*, 550–556. [CrossRef]
35. Varshowsaz, M.; Goorang, S.; Ehsani, S.; Azizi, Z.; Rahimian, S.J. Comparison of tissue density in Hounsfield units in computed tomography and cone beam computed tomography. *J. Dent.* **2016**, *13*, 108.
36. Arisan, V.; Karabuda, Z.C.; Avsever, H.; Özdemir, T. Conventional multi-slice computed tomography (CT) and cone-beam CT (CBCT) for computer-assisted implant placement. Part I: Relationship of radiographic gray density and implant stability. *Clin. Implant Dent. Relat. Res.* **2013**, *15*, 893–906. [CrossRef]
37. Liu, J.; Chen, H.-Y.; DoDo, H.; Yousef, H.; Firestone, A.R.; Chaudhry, J.; Johnston, W.M.; Lee, D.J.; Emam, H.A.; Kim, D.-G. Efficacy of cone-Beam computed tomography in evaluating bone quality for optimum implant treatment planning. *Implant Dent.* **2017**, *26*, 405–411. [CrossRef] [PubMed]
38. Monje, A.; Monje, F.; González-García, R.; Galindo-Moreno, P.; Rodriguez-Salvanes, F.; Wang, H.L. Comparison between microcomputed tomography and cone-beam computed tomography radiologic bone to assess atrophic posterior maxilla density and microarchitecture. *Clin. Oral Implants Res.* **2014**, *25*, 723–728. [CrossRef]
39. Naitoh, M.; Hirukawa, A.; Katsumata, A.; Ariji, E. Evaluation of voxel values in mandibular cancellous bone: Relationship between cone-beam computed tomography and multislice helical computed tomography. *Clin. Oral Implants Res.* **2009**, *20*, 503–506. [CrossRef]
40. Parsa, A.; Ibrahim, N.; Hassan, B.; van der Stelt, P.; Wismeijer, D.J. Bone quality evaluation at dental implant site using multislice CT, micro-CT, and cone beam CT. *Clin. Oral Implants Res.* **2015**, *26*, e1–e7. [CrossRef]
41. Nomura, Y.; Watanabe, H.; Honda, E.; Kurabayashi, T. Reliability of voxel values from cone-beam computed tomography for dental use in evaluating bone mineral density. *Clin. Oral Implants Res.* **2010**, *21*, 558–562. [CrossRef]
42. Baumgaertel, S. Quantitative investigation of palatal bone depth and cortical bone thickness for mini-implant placement in adults. *Am. J. Orthod. Dentofac. Orthop.* **2009**, *136*, 104–108. [CrossRef] [PubMed]
43. Baumgaertel, S.; Hans, M.G. Buccal cortical bone thickness for mini-implant placement. *Am. J. Orthod. Dentofac. Orthop.* **2009**, *136*, 230–235. [CrossRef] [PubMed]
44. Deguchi, T.; Nasu, M.; Murakami, K.; Yabuuchi, T.; Kamioka, H.; Takano-Yamamoto, T. Quantitative evaluation of cortical bone thickness with computed tomographic scanning for orthodontic implants. *Am. J. Orthod. Dentofac. Orthop.* **2006**, *129*, 721.e7–721.e12. [CrossRef] [PubMed]
45. Gerlach, N.L.; Meijer, G.J.; Borstlap, W.A.; Bronkhorst, E.M.; Bergé, S.J.; Maal, T.J.J. Accuracy of bone surface size and cortical layer thickness measurements using cone beam computerized tomography. *Clin. Oral Implants Res.* **2013**, *24*, 793–797. [CrossRef]
46. Ferrario, V.; Sforza, C.; Serrao, G.; Dellavia, C.; Tartaglia, G. Single tooth bite forces in healthy young adults. *J. Oral Rehabil.* **2004**, *31*, 18–22. [CrossRef]
47. Throckmorton, G.S.; Dean, J.S. The relationship between jaw-muscle mechanical advantage and activity levels during isometric bites in humans. *Arch. Oral Biol.* **1994**, *39*, 429–437. [CrossRef]
48. Cassetta, M.; Sofan, A.A.; Altieri, F.; Barbato, E. Evaluation of alveolar cortical bone thickness and density for orthodontic mini-implant placement. *J. Clin. Exp. Dent.* **2013**, *5*, e245. [CrossRef]

Article

Anatomical Variations of the Recurrent Laryngeal Nerve and Implications for Injury Prevention during Surgical Procedures of the Neck

Alison M. Thomas [1,2], Daniel K. Fahim [1,3] and Jickssa M. Gemechu [2,*]

1 Department of Neurosurgery, Oakland University William Beaumont School of Medicine, Rochester, MI 48309, USA; alison.thomas@beaumont.org (A.M.T.); Daniel.fahim@beaumont.org (D.K.F.)

2 Department of Foundational Medical Studies, Oakland University William Beaumont School of Medicine, Rochester, MI 48309, USA

3 Michigan Head & Spine Institute, Southfield, MI 48034, USA

* Correspondence: gemechu@oakland.edu; Tel.: +1-248-370-3667

Received: 18 August 2020; Accepted: 31 August 2020; Published: 4 September 2020

Abstract: Accurate knowledge of anatomical variations of the recurrent laryngeal nerve (RLN) provides information to prevent inadvertent intraoperative injury and ultimately guide best clinical and surgical practices. The present study aims to assess the potential anatomical variability of RLN pertaining to its course, branching pattern, and relationship to the inferior thyroid artery, which makes it vulnerable during surgical procedures of the neck. Fifty-five formalin-fixed cadavers were carefully dissected and examined, with the course of the RLN carefully evaluated and documented bilaterally. Our findings indicate that extra-laryngeal branches coming off the RLN on both the right and left side innervate the esophagus, trachea, and mainly intrinsic laryngeal muscles. On the right side, 89.1% of the cadavers demonstrated 2–5 extra-laryngeal branches. On the left, 74.6% of the cadavers demonstrated 2–3 extra-laryngeal branches. In relation to the inferior thyroid artery (ITA), 67.9% of right RLNs were located anteriorly, while 32.1% were located posteriorly. On the other hand, 32.1% of left RLNs were anterior to the ITA, while 67.9% were related posteriorly. On both sides, 3–5% of RLN crossed in between the branches of the ITA. Anatomical consideration of the variations in the course, branching pattern, and relationship of the RLNs is essential to minimize complications associated with surgical procedures of the neck, especially thyroidectomy and anterior cervical discectomy and fusion (ACDF) surgery. The information gained in this study emphasizes the need to preferentially utilize left-sided approaches for ACDF surgery whenever possible.

Keywords: recurrent laryngeal nerve; inferior thyroid artery; anatomical variations; ACDF; thyroidectomy

1. Introduction

The recurrent laryngeal nerve (RLN) is a branch of the vagus nerve (CN X). It carries sensory, motor, and parasympathetic fibers to the laryngeal structures [1]. It is the main motor nerve of all intrinsic laryngeal muscles, except the cricothyroid, which receives its innervation via the external laryngeal nerve [1]. The RLN has a different course on the left and right side of the body, where it loops around the arch of the aorta below the ligamentum arteriosum on the left side, and around the subclavian artery at the base of the neck on the right side [1]. On both sides, after looping deep to the relevant vessels, the RLN follows the tracheoesophageal groove to enter into the larynx to the inferior pharyngeal constrictor muscle [1,2].

It has been reported that anatomical variations of the RLN are commonly more rostral on the right than the left side, which is mainly explained due to the difference in their embryonic origin [1]. The left

RLN has a longer course in the thoracic region compared to the right RLN, which makes the former vulnerable to injury due to trauma and masses associated with oncologic diseases [3,4]. Although the right RLN travels within the tracheoesophageal groove, its course is more anterior and lateral than the left RLN, which causes many surgeons and researchers to debate the merits of left or right-sided approaches when performing anterior cervical discectomy and fusion (ACDF) procedures [3,5,6].

The safety of surgical procedures of the neck may be affected by potential anatomical variations of the RLN. In this regard, several variations have been reported in previous studies that make the nerve more vulnerable to damage [1,7]. Bilateral RLN variations in the same patient are also possible, which would increase the vulnerability of the nerve damage during surgical procedures of the neck [3]. Proper identification and preservation of the RLN and consideration of the potential variations may prevent or minimize the damage to the nerve.

Damage to the RLN has been reported as one of the most common iatrogenic complications associated with various surgeries of the neck [7]. RLN damage can be temporary or permanent, depending on the severity of the nerve injury and whether the injury is unilateral or bilateral. RLN damage results in hypophonation, and in extreme cases, dyspnea by paralyzing the muscles of the larynx [7]. Unilateral paralysis to the RLN may be clinically silent and present as hypophonia, dysphonia, or dysphagia and aspiration [8]. The fact that unilateral damage can be clinically silent results in RLN palsy being underdiagnosed and hence under-reported. This has the potential to render patients who undergo multiple procedures of the neck more vulnerable to bilateral damage. Bilateral damage to the RLN typically presents as dyspnea with inspiratory stridor, signifying a narrowed airway [2].

Lack of knowledge regarding the anatomical variations of RLN and its course and branching pattern carries potential risks, compromising the safety of surgical procedures in the neck. This cadaver-based study provides substantial information to close the knowledge gap and elucidate the relationship between the RLN and other anatomical structures in order to improve the safety of neck surgeries. Although previous studies have reported variations regarding the RLN based on a smaller number of cadavers, we endeavored to dissect a large number of cadavers to comprehensively investigate the variations of branching patterns, along with the potential risk of variations in the nerve's anatomical relationship with the inferior thyroid artery (ITA) [3,7–10].

2. Materials and Methods

The study was performed on 55 formalin-fixed cadavers at Oakland University William Beaumont School of Medicine from 2017–2019, of which 28 were male, 27 were female, and all were Caucasian in ethnicity. Anterior cervical deep dissection was performed in order to locate the RLN and surrounding landmarks on each cadaver. Cadavers that demonstrated anatomical variations were identified, and after detailed clearing, photographs were taken for data analysis. Cadavers without anatomical variations were also photographed for comparison as controls. The data was documented based on whether the left or right nerve was being observed, and also based on the number of branches that were identified on each RLN. We also took into account the relationship of the RLN with the ITA, and whether the nerve was anterior, posterior, or in between the branches of the artery. The data was analyzed quantitatively with the McNemar's test and Fisher's exact test, and statistical significance was determined based on a p value less than 0.05.

3. Results

The findings indicated that there were branches coming off the RLN on both the right and left sides, innervating both the esophagus and trachea in some instances. On the right side, 89.1% demonstrated anywhere between 2–5 extra-laryngeal branches, and 74.6% demonstrated branching on the left side (Table 1, Figure 1). Using the Fisher's exact test and subsequent data analysis, it was determined that there was a statistically significant difference in the branching pattern between the two sides. Of note,

16.4% of the cadavers demonstrated concomitant bilateral bifurcations in this study, while 16.4% of the cadavers demonstrated concomitant bilateral trifurcations.

Table 1. Frequency distribution of recurrent laryngeal nerve branching pattern.

	Right Side		**Left Side**	
	Frequency	**%**	**Frequency**	**%**
Branches		89.09		74.55
Bifurcation (2)	22	40.00	15	27.27
Trifurcation (3)	15	27.27	19	34.55
Multiple (≥4)	12	21.82	7	12.73
No branch (1)	6	10.91	14	25.45

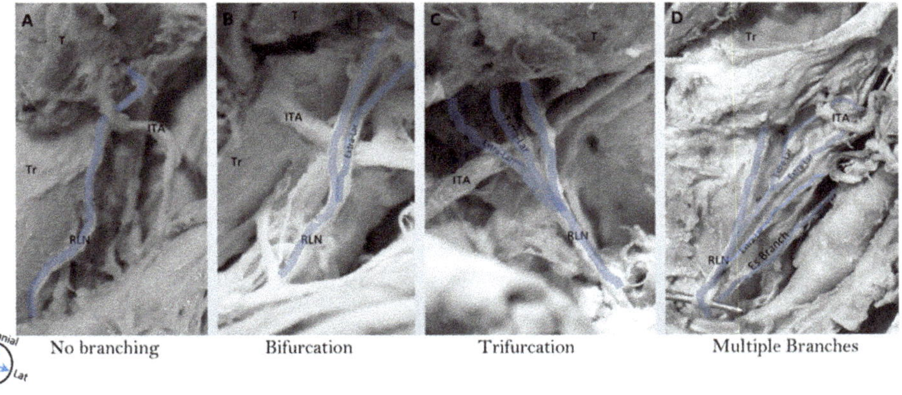

Figure 1. Extra-laryngeal branching pattern of recurrent laryngeal nerve. This figure demonstrates an example of the different branching patterns of the recurrent laryngeal nerve (RLN) documented in this study. (**A**) RLN without extra-laryngeal branching, (**B**) RLN with bifurcation (two branches), (**C**) RLN with trifurcation (three branches), and (**D**) RLN with multiple branches (four branches and above). Abbreviations: Es = esophagus, ITA = inferior thyroid artery, RLN = recurrent laryngeal nerve, T = thyroid, Tr = trachea.

Table 1 demonstrates the frequency distribution of branching patterns of the recurrent laryngeal nerve (RLN) on the right and left side. The right RLN branching pattern is documented to be anywhere between 2–5 branches (89.1%), while the left RLN is documented to be 2–5 branches (74.6%). The values were statistically significant, as determined by $p < 0.05$ (McNemar's test and Fisher's exact test, $p = 0.0348$).

The relationship of the RLN with the ITA was also examined. The ITA is a common landmark that surgeons use to locate the RLN, especially during thyroidectomy. The RLN was discovered to have a varying relationship with the ITA, and can be seen anterior or posterior to it, and even sometimes in between the branches of the ITA. In relation to the ITA, 67.9% of right RLNs were related anteriorly, while 32.1% were related posteriorly. The opposite values were true for the left side, with 67.9% of the RLNs related posteriorly and 32.1% related anteriorly. On the right side, 3.6% of nerves crossed in between branches of the ITA, while 5.4% of left RLNs were found crossing between branches of the ITA (Table 2, Figure 2).

Table 2. Frequency of relationship of recurrent laryngeal nerve to the inferior thyroid artery.

	Right Side		Left Side	
	Frequency	**%**	**Frequency**	**%**
Anterior	36	67.92	17	32.08
Posterior	17	32.08	36	67.92
In-between	2	3.57	3	5.36

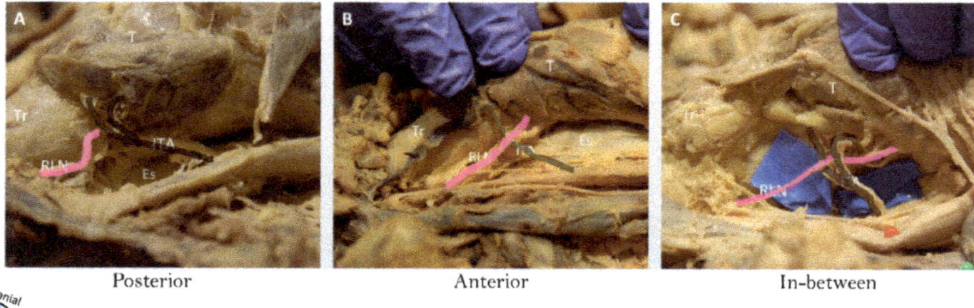

Figure 2. Relationship of recurrent laryngeal nerve to the inferior thyroid artery. This figure demonstrates the various relationships between the recurrent laryngeal nerve (RLN) and inferior thyroid artery (ITA). (**A**) demonstrates RLN related posterior to the ITA, (**B**) demonstrates RLN related anterior to the ITA, and (**C**) demonstrates RLN passing in between the two branches of the ITA. Abbreviations: Es = esophagus, ITA = inferior thyroid artery, RLN = recurrent laryngeal nerve, T = thyroid, Tr = trachea.

Table 2 demonstrates the relationship of the recurrent laryngeal nerve (RLN) to the inferior thyroid artery (ITA) on the right and left side. The RLN was observed as either related anteriorly, posteriorly, or in between the ITA. For the right RLN, 67.9% were anterior to the ITA, and 32.1% were posterior to it. The left RLN demonstrated the exact opposite results with 32.1% anterior, and 67.9% posterior to the ITA. The values demonstrated a statistically significant difference, as determined by $p < 0.05$ (McNemar's test and Fisher's exact test, $p = 0.0004$).

4. Discussion

Injury of the RLN is a well-known and troublesome complication associated with surgical procedures of the neck [9]. Awareness of anatomical variation in the branching patterns of the RLN contributes to the avoidance of this complication. Consideration of these variations is essential to minimize complications associated with surgical procedures of the neck—especially in ACDF and thyroidectomies. Significant postoperative complications, such as airway obstruction and narrowing, can be avoided if surgeons and their assistants in the operating room are aware of these variant structures. Lack of awareness of the anatomical variations in the branching pattern of the RLN makes it vulnerable to damage by stretching, compression, retraction, or accidental sharp division during the aforementioned surgical procedures.

Although a number of studies debate whether the left or right-sided approach to surgery is safer, there is no generally accepted consensus in the surgical community [5,6,8]. To answer this question, our study focused on observed side variations and found that the left RLN branched less often than the right RLN, and also took a more predictable course to the laryngeal structures, rather than an oblique, anterolateral approach. Chen and his colleagues found that overstretching of the RLN is less likely to occur on the left side due to the fact that it is better protected within the tracheoesophageal groove [11]. The right-sided approach is typically taken during an ACDF procedure due to surgeon handedness [11]. The other reason why neurosurgeons, in particular, favor a right-sided approach is due to the awareness of the risk of causing compression of the carotid artery during retraction or the inadvertent dislodging of atherosclerotic plaque during dissection, resulting in a stroke. A right-sided stroke is far better tolerated than a left-sided stroke due to language dominance in the overwhelming majority of patients.

In contrast to the thyroidectomy, in ACDF, the RLN is not routinely monitored or exposed as standard practice, potentially making it more prone to damage from indirect intraoperative injury by retraction or stretch injury while separating fascial layers or during retraction [12]. Of course, the counter-argument is that avoiding exposure of the nerve makes it less vulnerable to injury as no dissection is being carried out immediately around it. Direct surgical trauma to the RLN is rare, which is why injury is typically related to overstretching or excessive pressure from the endotracheal tube [13].

Our findings indicate that there is a significant amount of variation in the course and branching pattern of the RLN, indicating that 89% of right RLNs and 74.6% of left RLNs demonstrated 2–5 extra-laryngeal branches. These findings are in agreement with previous studies that have reported significant variability in RLN branching patterns [14–17]. This data suggests that the risk of iatrogenic injury is greater with right-sided approaches than left-sided approaches during ACDF procedures. The right RLN has more variability in its branching pattern, relationship with the ITA, and anterolateral position in comparison with the left [16,17]. Because of the fact that extra-laryngeal branching patterns are so common, misidentification of these branches can potentially lead to iatrogenic injury [16]. It is argued that branched nerves are even more vulnerable than those without branches because of their smaller caliber and fragility, ultimately making them prone to damage, even with normal manipulation [12]. Similar to our study, previous findings reported that 50% to 60% of patients often have small branches of the RLN that innervate the trachea, esophagus, or inferior constrictor muscles, and misidentification of these branches can cause a myriad of postoperative symptoms that include dysphagia, dyspnea, and dysphonia [16]. The greater number of branches on the right suggests that the risk of iatrogenic injury is greater with a right-sided approach than with a left-sided approach during ACDF procedures.

In addition to the greater number of branches, the relationship of the right RLN to the ITA may potentially make it more vulnerable to injury as well [16,17]. Campos and Henriques described a typical relationship between the RLN and ITA, where the RLN passed posterior to the nerve [17]. Our findings discredit the idea that the RLN consistently passes posterior to the ITA. Therefore, this idea of the RLN consistently passing posterior to the ITA should be a point of consideration, because it can give surgeons a false sense of security when dissecting anterior to ITA, ultimately leading to an increased risk of damage [13]. Our study, together with findings by other studies, provides evidence demonstrating that the right RLN more often passes anterior to the ITA approximately two-thirds of the time, while the inverse is true on the left [18]. Therefore, left-sided approaches to ACDF procedures may be safer as the ITA can more often be identified before the branches of the RLN are encountered.

In a meta-analysis study regarding anatomic variations of the RLN, a significant difference of 73.3% and 39.2% was noted in the prevalence of extra-laryngeal branch patterns between cadaveric studies and intraoperative studies, respectively [17]. The authors suggested that observed branching of the RLN was underestimated in the operating room, and this was attributed to the difficulty of viewing branches intraoperatively due to localized inflammation and edema that can be encountered in anterior neck dissections [18]. The true prevalence of extra-laryngeal branching was determined to be better reflected in cadaveric studies, prompting surgeons to attempt to expose the RLN and any branches in their entirety, unless it put the patient at risk for a more invasive procedure [17].

Vocal cord paralysis is one of the most significant morbidities after ACDF procedures, with incidence up to 24.2% immediately postoperatively [19]. Identification and localization of the RLN and intraoperative neuromonitoring of clinically relevant anatomical variations must be encouraged, along with the use of loupe magnification to identify the RLN and its landmarks [20]. Henry and others described how pre-operative ultrasound was used to identify structures and anatomical variations successfully [17]. This method successfully identified variants, such as nonrecurrent laryngeal nerves, 98% of the time, making it a reasonable method to decrease the risk of iatrogenic injury to the RLN [17]. Of course, this practice would be time-consuming and potentially cost prohibitive if undertaken for each patient prior to ACDF surgery. We recommend the use of ultrasound evaluation or intraoperative monitoring if a patient has had previous neck surgery. Alternatively, the patient may be subjected to direct laryngoscopy pre-operatively to evaluate for unilateral silent vocal cord paralysis. This population of patients may have clinically silent unilateral RLN injuries and are at higher risk for a potentially devastating injury if the contralateral side is injured in a subsequent procedure of the neck.

In general, the information gained from our study emphasizes the need for special considerations during ACDF and thyroidectomies, including side preferences, in order to preserve the extra-laryngeal branches of the RLN. Parameters that could potentially affect the incidence of branching patterns that must be evaluated further include variations in surgical exposure techniques, retraction practices, the use of surgical loops or magnification, and the use of intraoperative neuromonitoring [18]. The majority of previous studies performed assessing the anatomical variations and landmark structures of the RLN described only one aspect of the nerve and considered smaller sample sizes. Our study tried to address this drawback by considering a comprehensive anatomical approach to show the larger perspective of RLN and its relationship with the surrounding structures using a large sample size of cadavers. One of the limitations of this study is that all the assessed cadavers were Caucasian in ethnicity, and this might affect the generalizability of our findings.

In conclusion, this study may have implications for surgical technique and consideration of the side approach for preserving the extra-laryngeal branches of the RLN during surgical procedures of the neck such as ACDF and thyroidectomy procedures. In general, we recommend preferentially utilizing the left-sided approach for ACDF and being mindful of the proximity of the RLN branches to the ITA. Of course, the side of the surgical approach is multifactorial, and the individualized decision must be made by the surgeon.

Author Contributions: Conceptualization, A.M.T. and J.M.G.; data curation, A.M.T. and J.M.G.; formal analysis, A.M.T. and J.M.G.; investigation, A.M.T. and J.M.G.; methodology, A.M.T. and J.M.G.; project administration, A.M.T. and J.M.G.; resources, A.M.T., D.K.F. and J.M.G.; supervision J.M.G.; validation, A.M.T., D.K.F. and J.M.G.; visualization, A.M.T. and J.M.G.; writing—original draft, A.M.T.; writing—review & editing, A.M.T., D.K.F. and J.M.G. All authors have read and agreed to the published version of the manuscript.

Funding: This research received no external funding.

Acknowledgments: The Authors would like to acknowledge Mary Bee from the University of Detroit Mercy Lab and the OUWB Embark program staff for their unreserved support.

Conflicts of Interest: The authors declare no conflict of interest.

References

1. Shao, T.; Qiu, W.; Yang, W. Anatomical variations of the recurrent laryngeal nerve in Chinese patients: A prospective study of 2404 patients. *Sci. Rep.* **2016**, *6*, 25475. [CrossRef] [PubMed]
2. Kulekci, M.; Batıoglu-Karaaltın, A.; Saatci, O.; Uzun, I. Relationship between the Branches of the Recurrent Laryngeal Nerve and the Inferior Thyroid Artery. *Ann. Otol. Rhinol. Laryngol.* **2014**, *121*, 650–656. [CrossRef] [PubMed]
3. Haller, J.M.; Iwanik, M.; Shen, F.H. Clinically relevant anatomy of recurrent laryngeal nerve. *Spine* **2012**, *37*, 97–100. [CrossRef] [PubMed]
4. Dankbaar, J.W.; Pameijer, F.A. Vocal cord paralysis: Anatomy, imaging and pathology. *Insights Imaging* **2014**, *5*, 743–751. [CrossRef] [PubMed]
5. Miscusi, M.; Bellitti, A.; Peschillo, S.; Polli, F.M.; Missori, P.; Delfini, R. Does recurrent laryngeal nerve anatomy condition the choice of the side for approaching the anterior cervical spine? *J. Neurosurg. Sci.* **2007**, *51*, 61–64. [CrossRef] [PubMed]

6. Kilburg, C.; Sullivan, H.G.; Mathiason, M.A. Effect of approach side during anterior cervical discectomy and fusion on the incidence of recurrent laryngeal nerve injury. *J. Neurosurg. Spine* **2006**, *4*, 273–277. [CrossRef] [PubMed]
7. Cernea, C.R.; Hojaij, F.V.C.; De Carlucci, D.; Gotoda, R.; Plopper, C.; Vanderlei, F.; Brandão, L.G. Recurrent laryngeal nerve: A plexus rather than a nerve? *Arch. Otolaryngol. Head Neck Surg.* **2009**, *135*, 1098–1102. [CrossRef] [PubMed]
8. Jung, A.; Schramm, J. How to reduce recurrent laryngeal nerve palsy in anterior cervical spine surgery: A prospective observational study. *Neurosurgery* **2010**, *61*, 10–15. [CrossRef] [PubMed]
9. Liu, S.C.; Chou, Y.F.; Su, W.F. A rapid and Accurate Technique for the Identification of the Recurrent Laryngeal Nerve. *Ann. Otol. Rhinol. Laryngol.* **2014**, *123*, 805–810. [CrossRef] [PubMed]
10. Melamed, H.; Harris, M.B.; Awasthi, D. Anatomic Considerations of Superior Laryngeal Nerve During Anterior Cervical Spine Procedures. *Spine* **2002**, *27*, E83–E86. [CrossRef] [PubMed]
11. Chen, C.C.; Huang, Y.C.; Lee, S.T.; Chen, J.F.; Wu, C.T.; Tu, P.H. Long-term result of vocal cord paralysis after anterior cervical disectomy. *Eur. Spine J.* **2014**, *23*, 622–626. [CrossRef] [PubMed]
12. Rihn, J.A.; Kane, J.; Albert, T.J.; Vaccaro, A.R.; Hilibrand, A.S. What is the incidence and severity of dysphagia after anterior cervical surgery? *Clin. Orthop. Relat. Res.* **2011**, *469*, 658–665. [CrossRef] [PubMed]
13. Henry, B.M.; Vikse, J.; Graves, M.J.; Sanna, B.; Tomaszewska, I.M.; Tubbs, R.S.; Tomaszewski, K.A. Extralaryngeal branching of the recurrent laryngeal nerve: A meta-analysis of 28,387 nerves. *Langenbeck's Arch. Surg.* **2016**, *401*, 913–923. [CrossRef] [PubMed]
14. Gurleyik, E. Surgical anatomy of bilateral extralaryngeal bifurcation of the recurrent laryngeal nerve: Similarities and differences between both sides. *N. Am. J. Med. Sci.* **2014**, *6*, 445. [CrossRef] [PubMed]
15. Casella, C.; Pata, G.; Nascimbeni, R.; Mittempergher, F.; Salerni, B. Does extralaryngeal branching have an impact on the rate of postoperative transient or permanent recurrent laryngeal nerve palsy? *World J. Surg.* **2009**, *32*, 261. [CrossRef] [PubMed]
16. Uludag, M.; Yazici, P.; Aygun, N.; Citgez, B.; Yetkin, G.; Mihmanli, M.; Isgor, A. A Closer Look at the Recurrent Laryngeal Nerve Focusing on Branches & Diameters: A Prospective Cohort Study. *J. Investig. Surg.* **2016**, *29*, 383–388. [CrossRef]
17. Campos, B.A.; Henriques, P.R.F. Relationship between the recurrent laryngeal nerve and the inferior thyroid artery: A study in corpses. *Rev. Hosp. Clin.* **2000**, *55*, 195–200. [CrossRef] [PubMed]
18. Asgharpour, E.; Maranillo, E.; Sañudo, J.; Pascual-Font, A.; Rodriguez-Niedenführ, M.; Valderrama, F.J.; Viejo, F.; Parkin, I.G.; Vázquez, T. Recurrent laryngeal nerve landmarks revisited. *Head Neck* **2012**, *34*, 1240–1246. [CrossRef] [PubMed]
19. Tan, T.P.; Govindarajulu, A.P.; Massicotte, E.M.; Venkatraghavan, L. Vocal cord palsy after anterior cervical spine surgery: A qualitative systematic review. *Spine J.* **2014**, *14*, 1332–1342. [CrossRef] [PubMed]
20. D'Orazi, V.; Panunzi, A.; Di Lorenzo, E.; Ortensi, A.; Cialini, M.; Anichini, S.; Ortensi, A. Use of loupes magnification and microsurgical technique in thyroid surgery: Ten years experience in a single center. *G. Chir.* **2016**, *37*, 101–107. [CrossRef] [PubMed]

Article

Radiographic Analysis on the Distortion of the Anatomy of First Metatarsal Head in Dorsoplantar Projection

Jessica Grande-del-Arco [1], Ricardo Becerro-de-Bengoa-Vallejo [1], Patricia Palomo-López [2], Daniel López-López [3,*], César Calvo-Lobo [1], Eduardo Pérez-Boal [1], Marta Elena Losa-Iglesias [4], Carlos Martin-Villa [1] and David Rodriguez-Sanz [1]

1 Facultad de Enfermería, Fisioterapia y Podología, Universidad Complutense de Madrid, 28040 Madrid, Spain; jessicagrandedelarco@gmail.com (J.G.-d.-A.); ribebeva@ucm.es (R.B.-d.-B.-V.); cescalvo@ucm.es (C.C.-L.); perez.boal@gmail.com (E.P.-B.); podologiamartinvilla@gmail.com (C.M.-V.); davidrodriguezsanz@ucm.es (D.R.-S.)

2 University Center of Plasencia, Universidad de Extremadura, 10600 Plasencia, Spain; patibiom@unex.es

3 Research, Health and Podiatry Group, Department of Health Sciences, Faculty of Nursing and Podiatry, Universidade da Coruña, 15403 Ferrol, Spain

4 Faculty of Health Sciences, Universidad Rey Juan Carlos, 28922 Alcorcón, Spain; marta.losa@urjc.es

* Correspondence: daniel.lopez.lopez@udc.es; Tel.: +34-913-941-532

Received: 28 June 2020; Accepted: 31 July 2020; Published: 2 August 2020

Abstract: Background: The diagnostic of flat and crest-shaped of first metatarsal heads has been associated as an important risk factor for hallux deformities, such as hallux valgus and hallux rigidus. The rounded form of the first metatarsal head on the dorsoplantar radiograph of the foot has been believed to be associated with the development of hallux valgus. **Purpose:** The aim of this study was to clarify the effect of tube angulation on the distortion of first metatarsal head shape, and verify the real shape of the metatarsal head in anatomical dissection after an X-ray has been taken. **Materials and Methods:** In this prospective study at Universidad Complutense de Madrid, from December 2016 to June 2019, 103 feet from embalmed cadavers were included. We performed dorsoplantar radiograph tube angulation from 0° until 30° every 5° on all specimens; then, two observers verified the shape of the first metatarsal head in the radiographs and after its anatomic dissection. Kappa statistics and McNemar Bowker tests were used to assess and test for intra and interobserver agreement of metatarsal shape. **Results:** We calculated the intraobserver agreement, and the results showed that the first metatarsal head is distorted and crested only when the angle of the X-ray beam is at 20° of inclination ($p < 0.001$). The interobserver agreement showed good agreement at 0°, 5°, 10°, 20°, and 25° and was excellent at 30° ($p < 0.001$). **Conclusion:** All of the studies that we identified in the literature state that there are three types of shapes of the first metatarsal head and relate each type of head to the diagnosis of a foot pathology, such as hallux valgus or hallux rigidus. This study demonstrates that there is only the round-shaped form, and not three types of metatarsal head shape. Therefore, no diagnoses related to the shape of the first metatarsal head can be made.

Keywords: First metatarsal head; foot; radiological health; metatarsal bones

1. Introduction

Hallux valgus (HV) is a highly prevalent foot deformity estimated to affect 23% of adults and 35.7% of elderly individuals (1). HV presents a significant individual and public health burden, due to the high occurrence of related orthopedic foot surgery [1], and its association with foot pain [2,3], osteoarthritis (OA) at the first metatarsophalangeal joint (MTPJ), impaired gait patterns [4], poorly coordinated stability and an increased risk of falls in older adults [5,6].

While the development of HV is believed to be multifactorial, the exact etiology remains unclear [7]. Previous studies have suggested that several structural factors might be characteristic of HV, including various radiographic angles, first MTPJ congruency, metatarsal length, metatarsal head shape, sesamoid position, first metatarsocuneiform joint flexibility, and pes planus [8,9].

First metatarsal head shape has been routinely assessed by orthopedic surgeons radiographically, and has been addressed by as many as 24 authors, as well as in systematic reviews [7] to claim that shape is significant in the development of HV, and it has been classified as three types: round, square and crest [8,10–28], with the crest type being the most stable to prevent the development of HV and the round shape contributing to the development of HV, and it is one of the factors in recurrence after hallux valgus surgery [11,14,15,17–19,21–23,28–32].

Several authors have reported a relationship between a round-shaped metatarsal head and hallux valgus, but have not detected a strong correlation due to a lack of substantial data between them. Therefore, it is unknown whether a metatarsal head shape predisposes one to the development of hallux valgus.

In patients with HV, radiographs are obtained as part of a clinical evaluation. On these radiographs, angular measurement is used to determine the severity of deformation. A 1951 study [33] analyzed sources of error in the production and measurement of radiographs of the foot. This publication illustrated the need for the standardization of the radiograph of the dorsoplantar view of the foot, which has been widely advocated [33,34].

Despite this, various authors who described "their" standard technique of the dorsoplantar radiograph use a craniocaudal tube angulation of 5° [35], 15° [36–38], or 20° [39], but The American Orthopaedic Foot and Ankle Society recommended a tube angulation of 15° [40].

One study has been performed with a tube angulation of 20° in patients with HV, and states a relatively small reduction in the distortion of the intermetatarsal angles, but did not evaluate other anatomical structures.

To our knowledge, a systematic analysis of the relationship between tube angulation and the distortion due to the projection of the actual anatomy on the radiographs has not been performed beyond 20°.

The goal of this study was to analyze the effects and distortion that occur in the shape of the first metatarsal head when performing a dorsoplantar X-ray with the angled X-ray tube from 0° to 30° in anatomical specimens, and subsequently performing its dissection, to determine if the anatomic and radiographic findings correlate.

2. Material and Methods

From December 2016 to June 2019, 173 feet from embalmed cadavers were included in the study from Donation Center of the Bodies and Dissection Rooms of the Complutense, The University of Madrid. The institutional review board of the Rey Juan Carlos University approved with data 14 February of 2017 the study under number 27122011600917.

Those samples that included the complete foot with the distal third of the tibia and those samples that clinically showed no signs of surgical intervention were included in the selection of anatomical pieces.

The inclusion criteria followed in the radiographic evaluation were adult feet with radiographic images, in which all the growth cartilages of the foot and the distal third of the tibia and fibula were completely closed. It was required that the radiographic images showed the entire foot. Radiographs that showed traumatic or degenerative changes of the sesamoids or the surface of the first MTP joint, the presence of hallux valgus, or an intermetatarsal angle greater than 12° were excluded, as established in the article by Durrant et al. [41].

Each specimen was clinically examined to determine if they presented any deformity and those anatomical pieces that presented deformities in the foot, such as hallux valgus, hallux rigidus, osteoarthritis in the first MTF joint, fractures in the first metatarsal or presence of implants, patients with obviously abnormal shapes of the first metatarsal, due to fracture, invasion of the tumor, or congenital disease, were excluded.

Because of this, only 103 complying with the inclusion criteria were used outlined in the study.

The variables to be studied on the radiographs were the shape of the head of the 1st metatarsal, establishing the following categories: Round, square and "with crest", as reported in the literature [8,10–28].

2.1. Radiographic Protocol

The optimal tube angulation was defined as the angulation that was associated with the smallest average distortion. Besides the varying tube angulation, the geometry of this projection was identical to the standard technique of a dorsoplantar radiograph.

An Optima Xr200amx portable radiology equipment from Ge Heticare, 30 kW (GE HEALTHCARE, Madrid, Spain www.gehealthcare.com) was used with a 24 × 30 cm chassis and FireCR Spark Medical digital reader, 4dmedical, Valencia, Madrid.

The anatomical feet were placed on the radiographic plate in a neutral position, taking into account the methodology and protocol proposed by the studies by Venning and Hardy (1951) and Tanaka, Takakura, Kumai, Samotoy Tamai (1995) [33,42].

The standard dorsal, plantar radiographic projection proposed by several researchers was used: the X-ray beam tilts at 15° at a distance of 100 cm, to ensure the accuracy of these records are obtained from various articles [10,33,42].

2.2. Radiographic Representation

The samples underwent several images of the first metatarsal at different degrees of the beam projector. We perform a radiographic analysis with different degrees of projection.

We used a variable craniocaudal tube angulation in a sagittal plane 0°, 5°, 10°, 15°, 20°, 25°, and 30°, and the beam direction was set parallel to the axis of the foot and centered on the second metatarsal tarsus [42].

During X-ray imaging, the X-ray beam is perpendicular to the image intensifier, and the foot is positioned parallel to the image intensifier (Figure 1).

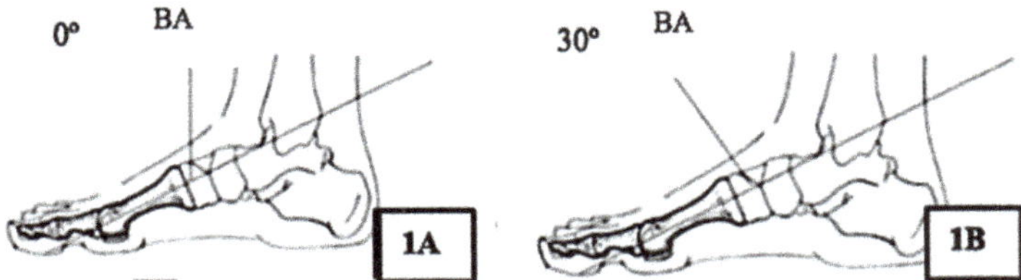

Figure 1. Position of the X-ray beam at a distance of 100 cm to obtain precision in the images. Abbreviations: BA: Beam angle. (**A**) relationship between the angulations with an X-ray beam projection at 0°; (**B**) Relationship between the angulations with an X-ray beam projection at 30°.

A neutral position, with 0° of inclination and rotation, avoiding pronation or supination of feet and beam direction focused on the second wedge joint as an exponent [43].

2.3. X-Ray Observation

The shape of the head of the first metatarsal was classified into three types, according to several authors: round, flat, and with crest [8,10–28]. The observation consisted of the two assessors measuring relevant measures of 103 randomly chosen feet radiographs, and then 1 week later re-measuring all radiographs without reference to previous results.

After observing the radiographs, the samples were dissected to assess the shape of the first metatarsal head by the same two observers who assessed the radiographs.

2.4. Statistical Analysis

Kappa statistics and generalized McNemar tests were used to assess and test for agreement. The shape of the first metatarsal head was polycotomized into three groups; "round", "flat", and "crest". As suggested by Landis and Koch, we interpreted the kappa values as follows: <0.20 indicates poor agreement, 0.21–0.40 fair, 0.41–0.60 moderate, 0.61–0.80 good agreement, and >0.80 indicates excellent agreement [44].

The McNemar Bowker test describes whether the marginal distributions of two measures are similar, as one would expect if the measures agree.

Data were analyzed using IBM SPSS Statistics, version 22 statistical software (SPSS Inc, Chicago, IL, USA). Statistical significance was set at $p < 0.05$, and Confidence Interval (IC) to 95%.

3. Results

The interobserver agreement by Kappa analysis (Table 1) showed a moderate agreement at 15°, good agreement at 0°, 5°, 10°, 20° and 25° and was excellent at 30°.

Table 1. Interobserver agreement about shape first metatarsal head in beam angle 0–30°.

			Reader B				
	Shape Metatarsal Head	Beam Angle	Round F (%)	Flat F (%)	Crest F (%)	Total F (%)	Kappa (p)
Reader A	0°	Round	69 (67.0%)	10 (9.7%)	0 (0.0%)	79 (76.7%)	0.679 (<0.001)
		Flat	3 (29%)	21 (20.4%)	0 (0.0%)	24 (23.3%)	
		Crest	0 (0.0%)	0 (0.0%)	0 (0.0%)	0 (0.0%)	
		Total F (%)	72 (69.9%)	32 (30.1%)	0 (0.0%)	103 (100%)	
	5°	Round	72 (69.9%)	9 (8.7%)	0 (0.0%)	81 (78.6%)	0.715 (<0.001)
		Flat	2 (1.9%)	20 (19.4%)	0 (0.0%)	22 (21.4%)	
		Crest	0 (0.0%)	0 (0.0%)	0 (0.0%)	0 (0.0%)	
		Total F (%)	74 (71.8%)	29 (28.2%)	0 (0.0%)	103 (100%)	
	10°	Round	73 (70.9%)	12 (11.7%)	0 (0.0%)	85 (82.5%)	0.613 (<0.001)
		Flat	2 (1.9%)	16 (15.5%)	0 (0.0%)	18 (17.5%)	
		Crest	0 (0.0%)	0 (0.0%)	0 (0.0%)	0 (0.0%)	
		Total F (%)	75 (72.8%)	28 (27.2%)	0 (0.0%)	103 (100%)	
	15°	Round	70 (68.0%)	10 (9.7%)	0 (0.0%)	80 (77.7%)	0.548 (<0.001)
		Flat	6 (5.8%)	16 (15.5%)	0 (0.0%)	22 (21.4%)	
		Crest	1 (0.0%)	0 (0.0%)	0 (0.0%)	1 (1.0%)	
		Total F (%)	75 (72.8%)	26 (27.2%)	0 (0.0%)	103 (100%)	
	20°	Round	38 (36.9%)	7 (6.8%)	2 (1.9%)	47 (45.6%)	0.726 (<0.001)
		Flat	1 (1.0%)	12 (11.7%)	0 (0.0%)	13 (12.6%)	
		Crest	2 (1.9%)	6 (5.8%)	35 (34.0%)	43 (42.7%)	
		Total F (%)	41 (39.8%)	25 (24.3%)	37 (35.9%)	103 (100%)	
	25°	Round	6 (5.8%)	1 (1.0%)	1 (1.0%)	8 (7.8%)	0.616 (<0.001)
		Flat	0 (0.0%)	3 (2.9%)	0 (0.0%)	3 (2.9%)	
		Crest	4 (3.9%)	4 (3.9%)	84 (81.6%)	92 (89.3%)	
		Total F (%)	10 (9.7%)	8 (7.8%)	85 (82.5%)	103 (100%)	
	30°	Round	2 (1.9%)	0 (0.0%)	0 (0.0%)	2 (1.9%)	0.853 (<0.001)
		Flat	0 (0.0%)	1 (1.0%)	0 (0.0%)	1 (1.0%)	
		Crest	0 (0.0%)	1 (1.0%)	99 (96.1%)	100 (97.1%)	
		Total F (%)	2 (1.9%)	2 (1.9%)	99 (96.1%)	103 (100%)	

Abbreviations: F, frequency.

To calculate intraobserver agreement, results were compared against angle beams. Table 2 shows the intraobserver A agreement regarding when the first metatarsal head gets distorted and appears crested. Results indicate that this occurs when the angle of the X-ray beam is at 20° of inclination. These results are similar for intraobserver B (Table 3), where the distortion of the same head occurs at 20° relative to 15° ($p < 0.001$).

Table 2. Intraobserver A agreement about shape first metatarsal head in beam angle: 0° vs. 5°, 5° vs. 10°, 10° vs. 15, 15 vs. 20°, 20 vs. 25°, 25° vs. 30°.

			Reader A				
	Shape Metatarsal Head	**Beam Angle**	**Round F (%)**	**Flat F (%)**	**Crest F (%)**	**Total F (%)**	**Mcnemar (*p* *)**
Reader A	0° vs. 5°	Round	79 (76.7)	0 (0.0%)	0 (0.0%)	79 (76.7%)	2 (0.157)
		Flat	2 (1.9%)	22 (21.4%)	0 (0.0%)	24 (23.3%)	
		Crest	0 (0.0%)	0 (0.0%)	0 (0.0%)	0 (0.0%)	
		Total F (%)	81 (78.6%)	22 (21.4%)	0 (0.0%)	103 (100%)	
	5° vs. 10°	Round	80 (77.7%)	1 (1.0%)	0 (0.0%)	81 (78.6%)	2.667 (0.102)
		Flat	5 (4.9%)	17 (16.5%)	0 (0.0%)	22 (21.4%)	
		Crest	0 (0.0%)	0 (0.0%)	0 (0.0%)	0 (0.0%)	
		Total F (%)	85 (82.5%)	18 (17.5%)	0 (0.0%)	103 (100%)	
	10° vs. 15°	Round	79 (76.7%)	5 (4.9%)	1 (0.0%)	85 (82.5%)	3.667 (0.160)
		Flat	1 (1.0%)	17 (16.5%)	0 (0.0%)	18 (17.5%)	
		Crest	1 (0.0%)	0 (0.0%)	0 (0.0%)	0 (0.0%)	
		Total F (%)	80 (77.7%)	22 (21.4%)	1 (1.0%)	103 (100%)	
	15° vs. 20°	Round	47 (45.6%)	1 (1.0%)	32 (31.1%)	80 (77.7%)	43 (<0.001)
		Flat	0 (0.0%)	12 (11.7%)	10 (9.7%)	22 (21.4%)	
		Crest	0 (0.0%)	0 (0.0%)	1 (1.0%)	1 (1.0%)	
		Total F (%)	47 (45.6%)	13 (12.6%)	43 (41.7%)	103 (100%)	
	20° vs. 25°	Round	7 (6.8%)	0 (0.0%)	40 (38.8%)	47 (45.6%)	50 (<0.001)
		Flat	1 (1.0%)	3 (2.9%)	9 (8.7%)	13 (12.6%)	
		Crest	0 (0.0%)	0 (0.0%)	43 (41.7%)	43 (41.7%)	
		Total F (%)	8 (7.8%)	3 (2.9%)	92 (89.3%)	103 (100%)	
	25° vs. 30°	Round	2 (1.9%)	0 (0.0%)	6 (5.8%)	8 (7.8%)	7 (0.030)
		Flat	0 (0.0%)	0 (0.0%)	3 (2.9%)	3 (2.9%)	
		Crest	0 (0.0%)	1 (1.0%)	91 (88.3%)	92 (89.3%)	
		Total F (%)	2 (1.9%)	1 (1.0%)	100 (97.5%)	103 (100%)	

Abbreviations: F: frequency. * The McNemar Bowker test was used to compare the relative prevalence of the different grades and is given as *p*-values.

Table 3. Intraobserver B agreement about shape first metatarsal head in beam angle: 0° vs. 5°, 5° vs. 10°, 10° vs. 15, 15 vs. 20°, 20 vs. 25°, 25° vs. 30°.

			Reader B				
	Shape Metatarsal head	**Beam Angle**	**Round F (%)**	**Flat F (%)**	**Crest F (%)**	**Total F (%)**	**Mcnemar (*p* *)**
Reader B	0° vs. 5°	Round	72 (69.9%)	0 (0.0%)	0 (0.0%)	72 (69.9%)	2 (0.157)
		Flat	2 (1.9%)	29 (28.2%)	0 (0.0%)	31 (30.1%)	
		Crest	0 (0.0%)	0 (0.0%)	0 (0.0%)	0 (0.0%)	
		Total F (%)	74 (71.8%)	29 (28.2%)	0 (0.0%)	103 (100%)	
	5° vs. 10°	Round	74 (71.8%)	0 (0.0%)	0 (0.0%)	74 (71.8%)	1 (0.317)
		Flat	1 (1.0%)	28 (27.2%)	0 (0.0%)	29 (28.2%)	
		Crest	0 (0.0%)	0 (0.0%)	0 (0.0%)	0 (0.0%)	
		Total F (%)	75 (72.8%)	28 (27.2%)	0 (0.0%)	103 (100%)	
	10° vs. 15°	Round	73 (70.9%)	2 (1.9%)	0 (0.0%)	75 (72.8%)	0.667 (0.414)
		Flat	4 (3.9%)	24 (23.3%)	0 (0.0%)	28 (27.2%)	
		Crest	0 (0.0%)	0 (0.0%)	0 (0.0%)	1 (0.0%)	
		Total F (%)	77 (74.8%)	26 (25.2%)	0 (0.0%)	103 (100%)	

Table 3. *Cont.*

			Reader B				
	Shape Metatarsal head	**Beam Angle**	**Round F (%)**	**Flat F (%)**	**Crest F (%)**	**Total F (%)**	**Mcnemar (p *)**
Reader B	15° vs. 20°	Round	41 (39.8%)	3 (2.9%)	33 (32.0%)	77 (74.8%)	39 (<0.001)
		Flat	0 (0.0%)	22 (21.4%)	4 (3.9%)	26 (25.2%)	
		Crest	0 (0.0%)	0 (0.0%)	0 (0.0%)	0 (0.0%)	
		Total F (%)	41 (39.8%)	25 (24.3%)	37 (35.9%)	103 (100%)	
	20° vs. 25°	Round	10 (9.7%)	4 (3.9%)	27 (26.2%)	41 (39.8%)	52 (<0.001)
		Flat	0 (0.0%)	4 (3.9%)	21 (20.4%)	25 (24.3%)	
		Crest	0 (0.0%)	0 (0.0%)	37 (35.9%)	37 (35.9%)	
		Total F (%)	10 (9.7%)	8 (7.8%)	85 (82.5%)	103 (100%)	
	25° vs. 30°	Round	2 (1.9%)	0 (0.0%)	8 (7.8%)	10 (9.7%)	14 (<0.001)
		Flat	0 (0.0%)	2 (1.9%)	6 (5.8%)	8 (7.8%)	
		Crest	0 (0.0%)	0 (0.0%)	85 (82.5%)	85 (82.5%)	
		Total F (%)	2 (1.9%)	2 (1.9%)	99 (96.1%)	103 (100%)	

Abbreviations: F, frequency. * The McNemar Bowker test was used to compare the relative prevalence of the different grades and is given as p-values.

Finally, after dissecting the 103 anatomical specimens, we found that all the first metatarsal heads had a round shape and none with a square shape or a crested head, showing perfect intra and interobserver agreement.

4. Discussion

The purpose of this study was to determine the presence of a distortion effect in the first metatarsal shape, due to the angulation of the X-ray beam.

Most articles on the measurement of dorsoplantar radiographs report a 15° [33,42] or 20° craniocaudal tube angulation. The American Orthopaedic Foot and Ankle Society has recommended a tube angulation of 15° [40].

We used observations of radiographs in this study. This technique used a craniocaudal tube angulation in a sagittal plane 0°, 5°, 10°, 15°, 20°, 25° and 30°, to evaluate the shape of the first metatarsal and including dissection of 103 feet embalmed cadaver by both observers.

We focused on the distorting effects of the tube angulation in the shape of the first metatarsal. We found that the distortion of the shape of the first metatarsal was minimal when the radiograph was made without angulation, or the beam angle was less than 20°.

Both observers agree that the shape of the metatarsal head is distorted in projections in which the X-ray beam with angulations is equal to or greater than 20° (Figure 2).

In this study, an association between a flat- or crested-shaped head of the first metatarsal with pathologies, such as hallux rigidus or hallux limitus, cannot be supported, because these shapes are the result of distortion caused by tube angulation.

Another reason for the distortion of the first metatarsal head with the X-ray beam correctly positions at 15° is that the normal first metatarsal declination angle is 21° angle between the axis of the first metatarsal and a horizontal linear [45], and in the flat foot, the first metatarsal declination angle is lower.

So, we postulated that when tube angulation in a sagittal plane is 15° [33,42] in a normal foot with a first metatarsal angle declination of 21°, the possibility of deformation or distortion of the first metatarsal head is minimized.

Instead, if the first metatarsal bone is dorsiflexed as a flat foot, the first metatarsal angle declination is lower, and the angle between the X-ray beam and the axis of the first metatarsal bone is a higher, thus maximizing distortion of the first metatarsal head.

In light of these findings, it seems necessary to control the beam angulation to 5–10° in dorsoplantar X-rays of the flat loading foot, to avoid the presence of the crested or flat shape, which are artifacts produced by the angulation of the tube.

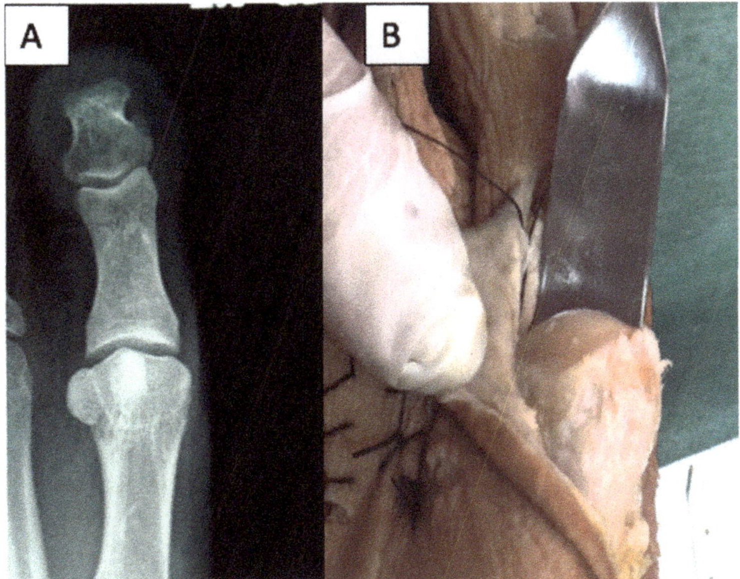

Figure 2. Views of a first metatarsal head showing distortion to appear crest shaped in a radiographic image performed to 30° (**A**) and after dissection revealing a round shape (**B**).

5. Conclusions

All of the articles that we identified state that there are three types of shapes of the first metatarsal head, and all authors relate each type of head to the diagnosis of a foot pathology, such as hallux valgus or hallux rigidus. This study demonstrates that there is only a round shape, and not three types of metatarsal head shape, and therefore, no diagnoses related to the shape of the first metatarsal head can be made.

A clinician should be aware that, in patients with flat feet, dorsoplantar with weight projection should be taken at an angle of the 5 to 10° beam.

Author Contributions: All authors: concept, design, analyses, interpretation of data, drafting of manuscript or revising it critically for important intellectual content. All authors have read and agreed to the published version of the manuscript.

Funding: This research received no external funding.

Conflicts of Interest: The authors declare no conflict of interest

References

1. Meyr, A.J.; Adams, M.L.; Sheridan, M.J.; Ahalt, R.G. Epidemiological Aspects of the Surgical Correction of Structural Forefoot Pathology. *J. Foot Ankle Surg.* **2009**, *48*, 543–551. [CrossRef] [PubMed]
2. Abhishek, A.; Roddy, E.; Zhang, W.; Doherty, M. Are hallux valgus and big toe pain associated with impaired quality of life? A cross-sectional study. *Osteoarthr. Cartil.* **2010**, *18*, 923–926. [CrossRef] [PubMed]
3. Roddy, E.; Zhang, W.; Doherty, M. Prevalence and associations of hallux valgus in a primary care population. *Arthritis Care Res.* **2008**, *59*, 857–862. [CrossRef] [PubMed]

4. Menz, H.B.; Lord, S.R. Gait instability in older people with hallux valgus. *Foot Ankle Int.* **2005**, *26*, 483–489. [CrossRef] [PubMed]
5. Koski, K.; Luukinen, H.; Laippala, P.; Kivelä, S.L. Physiological factors and medications as predictors of injurious falls by elderly people: A prospective population-based study. *Age Ageing* **1996**, *25*, 29–38. [CrossRef]
6. Spink, M.J.; Fotoohabadi, M.R.; Wee, E.; Hill, K.D.; Lord, S.R.; Menz, H.B. Foot and ankle strength, range of motion, posture, and deformity are associated with balance and functional ability in older adults. *Arch. Phys. Med. Rehabil.* **2011**, *92*, 68–75. [CrossRef]
7. Nix, S.E.; Vicenzino, B.T.; Collins, N.J.; Smith, M.D. Characteristics of foot structure and footwear associated with hallux valgus: A systematic review. *Osteoarthr. Cartil.* **2012**, *20*, 1059–1074. [CrossRef]
8. Coughlin, M.J.; Jones, C.P. Hallux Valgus: Demographics, Etiology, and Radiographic Assessment. *Foot Ankle Int.* **2007**, *28*, 759–777. [CrossRef]
9. Hardy, R.H.; Clapham, J.C. Observations on hallux valgus; based on a controlled series. *J. Bone Jt. Surg. Br.* **1951**, *33B*, 376–391. [CrossRef]
10. Ferrari, J.; Malone-Lee, J. The shape of the metatarsal head as a cause of hallux abductovalgus. *Foot Ankle Int.* **2002**, *23*, 236–242. [CrossRef]
11. Mann, R.; Coughlin, M.J. Adult hallux valgus. In *Surgery of the Foot and Ankle*; Mosby: St. Louis, MO, USA, 1999; pp. 150–269.
12. Van Deventer, S.J.; Strydom, A.; Saragas, N.P.; Ferrao, P.N.F. Morphology of the first metatarsal head as a risk factor for hallux valgus interphalangeus. *Foot Ankle Surg.* **2018**, *26*, 105–109. [CrossRef] [PubMed]
13. Beeson, P.; Phillips, C.; Corr, S.; Ribbans, W.J. Cross-sectional study to evaluate radiological parameters in hallux rigidus. *Foot* **2009**, *19*, 7–21. [CrossRef] [PubMed]
14. Okuda, R.; Kinoshita, M.; Yasuda, T.; Jotoku, T.; Kitano, N.; Shima, H. The shape of the lateral edge of the first metatarsal head as a risk factor for recurrence of hallux valgus. *J. Bone Jt. Surg. Am.* **2007**, *89*, 2163–2172. [CrossRef]
15. Burns, P.R.; Mecham, B. Biodynamics of Hallux Abductovalgus Etiology and Preoperative Evaluation. *Clin. Podiatr. Med. Surg.* **2014**, *31*, 197–212. [CrossRef]
16. Kilmartin, T.E.; Wallace, W.A. First metatarsal head shape in juvenile hallux abducto valgus. *J. Foot Surg.* **1991**, *30*, 506–508.
17. Palladino, S. Preoperative evaluation of the bunion patient: Etiology, biomechanics, clinical and radiographic assessment. In *Textbook of Bunion Surgery*, 2nd ed.; Gerbert, J., Ed.; Futura Publishing Co.: Mt. Kisco, NY, USA, 1991; pp. 1–88.
18. Brahm, S. Shape of the first metatarsal head in hallux rigidus and hallux valgus. *J. Am. Podiatr. Med. Assoc.* **1988**, *78*, 300–304. [CrossRef]
19. Laporta, D.M.; Melillo, T.V.; Hetherington, V.J. Preoperative Assessment in Hallux Valgus. *Hallux Valgus Forefoot Surg.* **2002**, 107–123. Available online: https://www.kent.edu/sites/default/files/HV-ch-06-Preoperative-Assessment-in-Hallux-Valgus.pdf (accessed on 24 January 2019).
20. Michelson, J.D.; Janowski, J.W.; Charlson, M.D. Quantitative relationship of first metatarsophalangeal head morphology to hallux rigidus and hallux valgus. *Foot Ankle Surg.* **2018**, *24*, 435–439. [CrossRef]
21. ElSaid, A.G.; Tisdel, C.; Donley, B.; Sferra, J.; Neth, D.; Davis, B. First Metatarsal Bone: An Anatomic Study. *Foot Ankle Int.* **2006**, *27*, 1041–1048. [CrossRef]
22. Duvries, H.L. Static deformities. In *Surgery of the Foot*; Mosby: St. Louis, MO, USA, 1959; pp. 346–442.
23. Karasick, D.; Wapner, K.L. Hallux valgus deformity: Preoperative radiologic assessment. *Am. J. Roentgenol.* **1990**, *155*, 119–123. [CrossRef]
24. Mancuso, J.E.; Abramow, S.P.; Landsman, M.J.; Waldman, M.; Carioscia, M. The zero-plus first metatarsal and its relationship to bunion deformity. *J. Foot Ankle Surg.* **2003**, *42*, 319–326. [CrossRef] [PubMed]
25. De Pablos, J.M.; Gómez, B.S.; Sabaté, D.J.; Del Boz, M.J.; Vázquez, J. Factores predisponentes del Hallux Valgus: Valoración Radiológica. *Rev. Med. Cir. Pie.* **1995**, *9*, 21–26.
26. Gutiérrez, C.P.; Sebastián, F.E.; Beltoldi, L.G. Factores morfológicos que influyen en el hallux valgus. *Rev. Esp. Cir. Ortop. Traumatol.* **1998**, *42*, 356–362.
27. Fellner, D.; Milsom, P. Relationship between hallux valgus and first metatarsal head shape. *J. Br. Podiatr Med.* **1995**, *50*, 54–56.

28. Martin, D.E.; Pontious, J. Introduction and Evaluation of Hallux Abducto Valgus. In *McGlamry's Comprehensive Textbook of Foot and Ankle Surgery*; Wolters Kluwer Health: Pennsylvania, PA, USA, 2012.
29. Laporta, G.; Melillo, T.; Olinsky, D. X-ray evaluation of hallux abducto valgus deformity. *J. Am. Podiatr. Med. Assoc.* **1974**, *64*, 544–566. [CrossRef]
30. Coughlin, M.J. Hallux valgus. *Instr. Course Lect.* **1997**, *46*, 357–391. [CrossRef]
31. Coughlin, M.J.; Shurnas, P.S. Hallux rigidus: Demographics, etiology, and radiographic assessment. *Foot Ankle Int.* **2003**, *24*, 731–743. [CrossRef]
32. Coughlin, M.J.; Saltzman, S.L.; Anderson, R.B. Hallux valgus. In *Mann's Surgery of the Foot and Ankle*; Saunders/Elsevier: Philadelphia, PA, USA, 2007; pp. 183–184.
33. Venning, P.; Hardy, R.H. Sources of Error in the Production and Measurement of Standard Radiographs of the Foot. *Br. J. Radiol.* **1951**, *24*, 18–26. [CrossRef]
34. Stevens, P. Radiographic distortion of bones: A marker study. *Orthopedics* **1989**, *12*, 1457–1463.
35. Shereff, M.J.; DiGiovanni, L.; Bejjani, F.J.; Hersh, A.; Kummer, F. A comparison of nonweight-bearing and weight-bearing radiographs of the foot. *Foot Ankle* **1990**, *10*, 306–311. [CrossRef]
36. Hlavac, H.F. Differences in x-ray findings with varied positioning of the foot. *J. Am. Podiatry Assoc.* **1967**, *57*, 465–471. [CrossRef] [PubMed]
37. Kaschak, T.J.; Laine, W. Surgical radiology. *Clin. Podiatr. Med. Surg.* **1988**, *5*, 797–829. [PubMed]
38. Tanaka, Y.; Takakura, Y.; Takaoka, T.; Akiyama, K.; Fujii, T.; Tamai, S. Radiographic analysis of hallux valgus in women on weightbearing and nonweightbearing. *Clin. Orthop. Relat. Res.* **1997**, 186–194. [CrossRef] [PubMed]
39. Saltzman, C.L.; Brandser, E.A.; Berbaum, K.S.; Degnore, L.; Holmes, J.R.; Katcherian, D.A.; Teasdall, R.D.; Alexander, I.J. Reliability of Standard Foot Radiographic Measurements. *Foot Ankle Int.* **1994**, *15*, 661–665. [CrossRef] [PubMed]
40. Smith, R.W.; Reynolds, J.C.; Stewart, M.J. Hallux valgus assessment: Report of research committee of American Orthopaedic Foot and Ankle Society. *Foot Ankle* **1984**, *5*, 92–103. [CrossRef] [PubMed]
41. Durrant, M.; McElroy, T. Radiographic image distortion between the distal edge of the first metatarsal and the tibial sesamoid: Establishing a reliable radiographic relationship. *J. Am. Podiatr. Med. Assoc.* **2010**, *100*, 1–9. [CrossRef]
42. Tanaka, Y.; Takakura, Y.; Kumai, T.; Samoto, N.; Tamai, S. Radiographic analysis of hallux valgus. A two-dimensional coordinate system. *J. Bone Jt. Surg.* **1995**, *77*, 205–213. [CrossRef]
43. Tanaka, Y.; Takakura, Y.; Fujii, T.; Kumai, T.; Sugimoto, K. Hindfoot alignment of hallux valgus evaluated by a weightbearing subtalar x-ray view. *Foot Ankle Int.* **1999**, *20*, 640–645. [CrossRef]
44. Landis, J.R.; Koch, G.G. The measurement of observer agreement for categorical data. *Biometrics* **1977**, *33*, 159–174. [CrossRef]
45. Mark Davies, A.; Whitehouse, R.; Jenkins, J. *Imaging of the Foot & Ankle: Techniques and Applications*; Springer: Berlin/Heidelberg, Germany, 2003.

Case Report

Frequency and Clinical Review of the Aberrant Obturator Artery: A Cadaveric Study

Guinevere Granite [1,*], Keiko Meshida [2] and Gary Wind [1]

[1] Department of Surgery, Uniformed Services University of the Health Sciences, Bethesda, MD 20814, USA; gary.wind@usuhs.edu

[2] Department of Anatomy, College of Medicine, Howard University, Washington, DC 20059, USA; keiko.meshida@howard.edu

* Correspondence: guinevere.granite@usuhs.edu; Tel.: +1-301-295-1500

Received: 12 June 2020; Accepted: 29 July 2020; Published: 30 July 2020

Abstract: The occurrence of an aberrant obturator artery is common in human anatomy. Detailed knowledge of this anatomical variation is important for the outcome of pelvic and groin surgeries requiring appropriate ligation. Familiarity with the occurrence of an aberrant obturator artery is equally important for instructors teaching pelvic anatomy to students. Case studies highlighting this vascular variation provide anatomical instructors and surgeons with accurate information on how to identify such variants and their prevalence. Seven out of eighteen individuals studied (38.9%) exhibited an aberrant obturator artery, with two of those individuals presenting with bilateral aberrant obturator arteries (11.1%). Six of these individuals had an aberrant obturator artery that originated from the deep inferior epigastric artery (33.3%). One individual had an aberrant obturator artery that originated directly from the external iliac artery (5.6%).

Keywords: aberrant obturator artery; internal iliac branching variations; external iliac branching variations; anatomical variations

1. Introduction

Knowledge of the vascular anatomy and its possible variations is essential to performing embolization and revascularization procedures in the human pelvis [1]. The obturator artery (OA), a standard branch of the anterior division of the internal iliac artery (IIA), has the greatest frequency of variation among the IIA branches [2–12]. Understanding the possible OA origin variations is important while performing pelvic and groin surgeries requiring appropriate ligation. In instances of acute pelvic or groin trauma, such varying origins may be a significant source of persistent hemorrhages that are difficult to manage [10,11,13–25]. Thus, OA origin variations are an important anatomical topic for a range of medical fields as varied as gynecology, orthopedics, and urology [25,26].

The abdominal aorta divides into the right and left common iliac arteries, which further subdivide into the external iliac artery (EIA) and the internal iliac artery (IIA) on each side [11,13,27,28]. The EIA mainly supplies the lower limbs. The IIA usually supplies the pelvis, perineum, and gluteal regions with common anatomical variations [1,11,23,26,28]. Typically, four IIA branches occur in the male, while five occur in the female (see Section 3.1 for more details) [1,11,23,26,28].

An aberrant obturator artery (AOA) is an anatomical variation in which the OA often arises from the external iliac artery (EIA) (Figure 1) [3,21,27,28]. Select case studies have identified it in as many as 55.1% of individuals in their cohort [3,4,21,27,28]. Other alternative OA origins include the common iliac artery (CIA), inferior gluteal artery (IGA), internal pudendal artery (IPA), a common trunk for IGA and IPA, iliolumbar artery (ILA), EIA, a branch of the EIA, or by a dual root from both IIA and EIA sources [2,3,5,7,9–11,14,21,23,27–33].

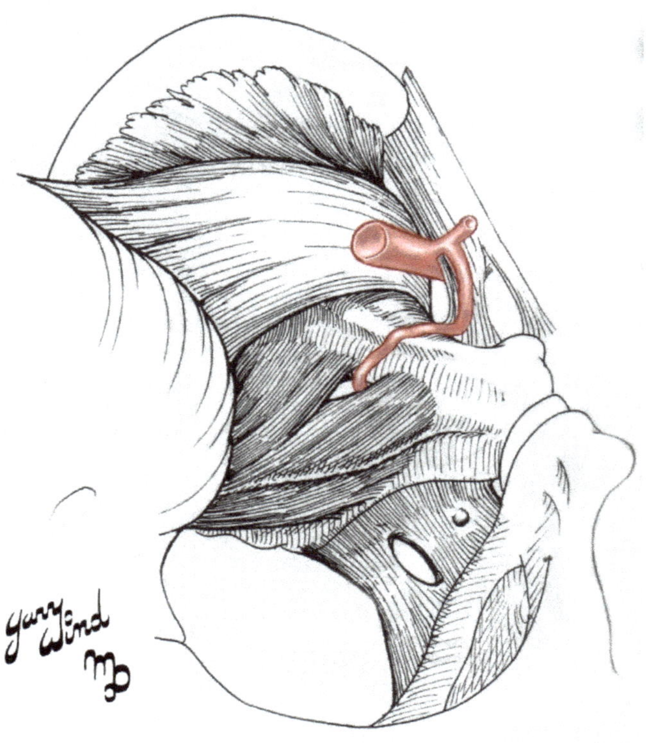

Figure 1. Illustrative schematic of the most common type of the aberrant obturator artery.

Familiarity with the occurrence of an AOA is equally important for instructors teaching pelvic anatomy to students. Case studies highlighting this vascular variation provide anatomical instructors and surgeons with accurate information on how to identify such variants and their prevalence. In our studied population, the OA arose from the IIA bilaterally in ten of the eighteen individuals (55.6%). The OA branched from the posterior division unilaterally in two cadavers (11.1%); one on the left and one on the right. The individual with an OA originating from the left IIA posterior division had a right obturator artery (ROA) arising from the right IIA anterior division. The individual with an OA arising from the right IIA posterior division also had a left AOA (LAOA). Seven of the eighteen studied individuals (38.9%) had at least one AOA. Two cadavers had bilateral AOAs (11.1%). The cadavers analyzed were provided by the Maryland State Anatomy Board and the Uniformed Services University of the Health Sciences Donation Program.

A thorough understanding of the IIA branching patterns and their possible vascular variations is essential for obstetric surgeons, general surgeons, and interventional radiologists performing other types of pelvic procedures (i.e., hernia repairs or pelvic fractures) (see Section 3.4 for more details) [3,10,11,19,23,27,29,32,34].

2. Case Descriptions

Case 1: A 96-year-old White Male with a listed cause of death of dementia exhibited a unilateral AOA on the left pelvic side with a Yamaki et al. (1998) Group B classification bilaterally and a Sañudo et al. (2011) Type B classification (Tables 1 and 2) [8,26]. The left IIA posterior division branches (LIIA PDB) included the left iliolumbar artery (LILA), left superior gluteal artery (LSGA), left superior lateral sacral artery (LSLSA), left inferior lateral sacral artery (LILSA), and the left inferior gluteal artery (LIGA). The left IIA anterior division branches (LIIA ADB) were the left umbilical artery (LUA), with a branch of the left superior vesical artery (LSVA); the LSVA, with a branch of the left inferior vesical artery (LIVA); and the left prostatic artery (LPA), which has a common branch that bifurcated into the left internal pudendal artery (LIPA) and the left middle rectal artery (LMRA). The left aberrant obturator artery (LAOA) originated from the left deep inferior epigastric artery (LDIEA), which was a branch of the left external iliac artery (LEIA) (Figure 2). The right IIA posterior division branches (RIIA PDB) included the right iliolumbar artery (RILA), the right superior lateral sacral artery (RSLSA), the right inferior lateral sacral artery (RILSA), the right superior gluteal artery (RSGA), and the right inferior gluteal artery (RIGA). The right IIA anterior division branched (RIIA ADB) into the right umbilical artery (RUA), with a branch for the right superior vesical artery (RSVA); the right obturator artery (ROA); and a common branch for the right inferior vesical artery (RIVA), the right internal pudendal artery (RIPA), and the right middle rectal artery (RMRA). The LIIA PDB had a left accessory obturator artery branch (LAOAB) providing multiple arterial branches to the left iliacus and iliopsoas muscles. The ROA provided multiple arterial branches to these muscles on the right side. There was no distinct separation between the anterior and posterior divisions on either side of the pelvis.

Table 1. Yamaki et al. (1998) [26] IIA branching pattern classification system. SGA = superior gluteal artery; IGA = inferior gluteal artery; IPA = internal pudendal artery.

Group	Description of IIA Divisions
A	Two branches: (1) SGA and (2) a common trunk of IGA and IPA (common gluteal-pudendal trunk)
B	Two branches: (1) IPA and (2) a common gluteal trunk of SGA and IGA
C	Three branches: (1) SGA, (2) IGA, and (3) IPA
D	Two branches: (1) a common trunk of SGA and IPA and (2) IGA

Table 2. Sañudo et al. (2011) [8] obturator artery variation types. OA = obturator artery; DIEA = deep inferior epigastric artery; IIA = internal iliac artery; EIA = external iliac artery; FA = femoral artery.

Type	Description of OA Variation
A	The OA arises from anterior division of the IIA
B	The OA arises from the DIEA
C	The OA is a branch of the posterior division of the IIA
D	The OA arises from the IIA, above its final branching
E	The OA arises from the EIA
F	The OA arises from the FA

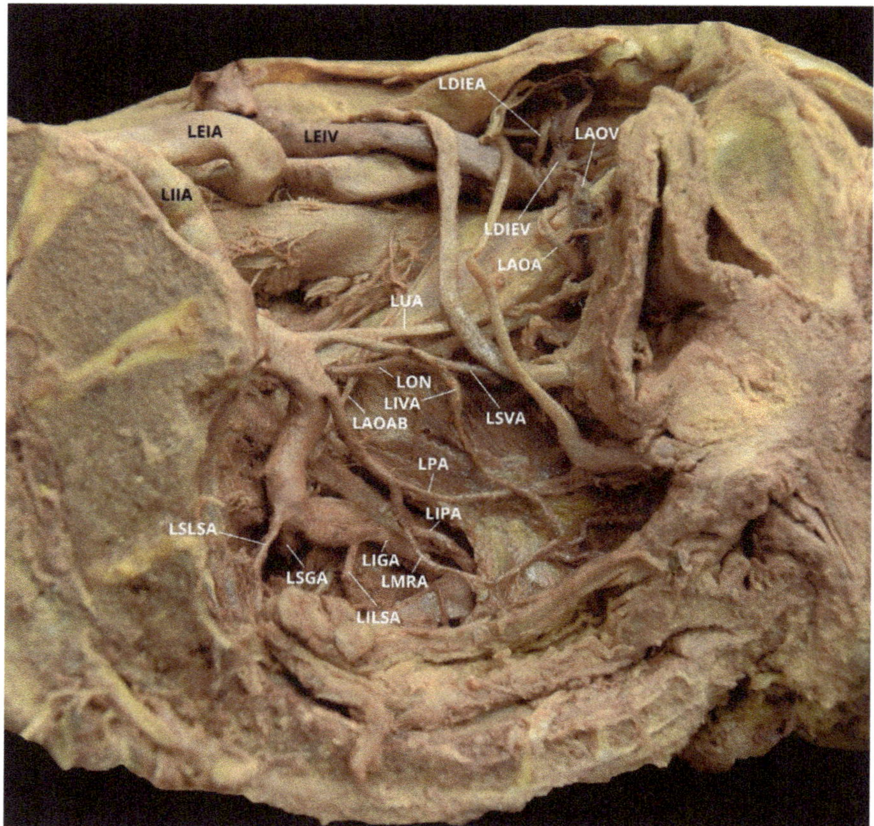

Figure 2. Facilitated display highlighting the left internal iliac artery branches with the left aberrant obturator artery for Case 1. LAOA = Left Aberrant Obturator Artery; LAOAB = Left Accessory Obturator Artery Branch; LAOV = Left Aberrant Obturator Vein; LDIEA = Left Deep Inferior Epigastric Artery; LDIEV = Left Deep Inferior Epigastric Vein; LEIA = Left External Iliac Artery; LEIV = Left External Iliac Vein; LIGA = Left Inferior Gluteal Artery; LIIA = Left Internal Iliac Artery; LILSA = Left Inferior Lateral Sacral Artery; LIPA = Left Internal Pudendal Artery; LIVA = Left Inferior Vesical Artery; LMRA = Left Middle Rectal Artery; LON = Left Obturator Nerve; LPA = Left Prostatic Artery; LSGA = Left Superior Gluteal Artery; LSLSA = Left Superior Lateral Sacral Artery; LSVA = Left Superior Vesical Artery; LUA = Left Umbilical Artery. The Left Iliolumbar Artery (LILA) is not pictured.

Case 2: A 70-year-old White Male with a listed cause of death of cardiac arrest exhibited bilateral AOAs with a Yamaki et al. (1998) Group B classification on the left, a Group A classification on the right, and a Sañudo et al. (2011) Type B classification bilaterally (Tables 1 and 2) [8,26]. The LIIA PDB included the LILA and the LSGA with a terminating branch of LIGA. The LIIA ADB were the LUA, a possible LSVA, a possible LIVA, and the LIPA (Figure 3). The RIIA PDB were the RLSLA and the RSGA. The RIIA ADB were the RUA, with a branch of RSVA; the right middle vesical (or vesiculodeferential) artery (RMVA); the RIVA; and the RIGA with a branch of the RMRA and the RIPA (Figure 4). There was a LAOA and a right aberrant obturator artery (RAOA) with this individual in which both originated from their respective DIEAs. Due to damage that occurred during student dissection, the LSLSA, LILSA, LMRA, RILA, and the RILSA were not identifiable and several branches could not be named with certainty.

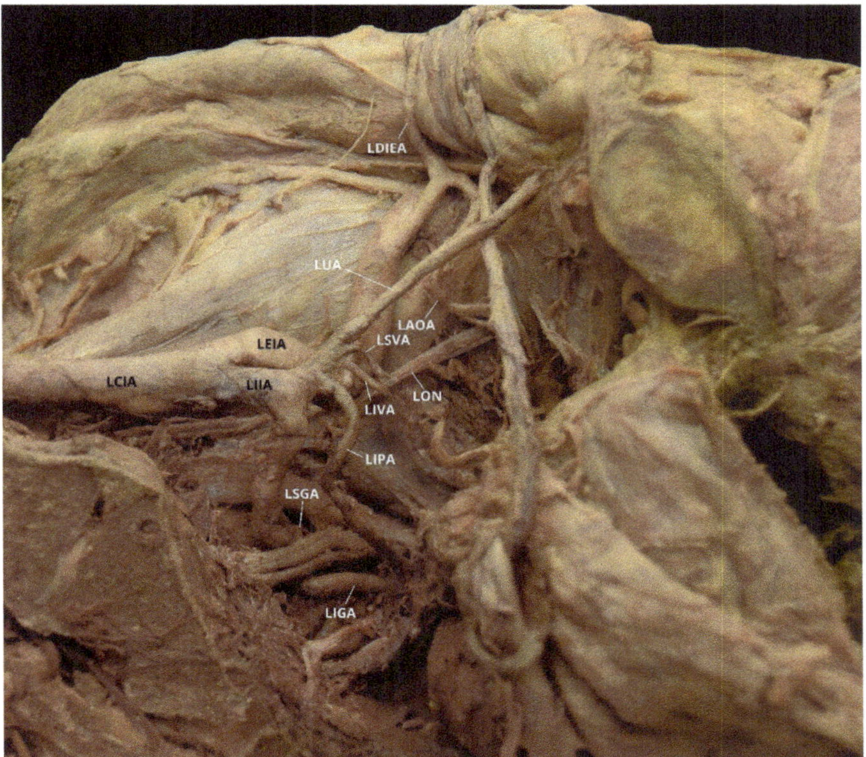

Figure 3. Facilitated display highlighting the left internal iliac artery branches with the left aberrant obturator artery for Case 2. LAOA = Left Aberrant Obturator Artery; LCIA = Left Common Iliac Artery; LDIEA = Left Deep Inferior Epigastric Artery; LEIA = Left External Iliac Artery; LIGA = Left Inferior Gluteal Artery; LIIA = Left Internal Iliac Artery; LIPA = Left Internal Pudendal Artery; LIVA = Left Inferior Vesical Artery; LON = Left Obturator Nerve; LSGA = Left Superior Gluteal Artery; LSVA = Left Superior Vesical Artery; LUA = Left Umbilical Artery. The Left Iliolumbar Artery (LILA), Left Superior Lateral Sacral Artery (LSLSA), Left Inferior Lateral Sacral Artery (LILSA), and the Left Middle Rectal Artery (LMRA) are not pictured.

Case 3: An 84-year-old White Female with a listed cause of death of senile degeneration of brain/Parkinson's disease was found to have a unilateral AOA on the left pelvic side with a Yamaki et al. (1998) Group A classification bilaterally and a Sañudo et al. (2011) Type B classification (Tables 1 and 2) [8,26]. The LIIA PDB were the LILA, the LSLSA, the LILSA, and the LSGA. The LIIA ADB included the LUA; the LSVA, with a branch of LUTA; the LIPA, with the left vaginal artery (LVA) branch and the LMRA branch; and the LIGA. The LAOA originated from the LDIEA (Figure 5). The RIIA PDB were the RILA, the RLSA, and the RSGA. The RIIA ADB were as follows: the RUA, with a RSVA branch that had a right uterine artery (RUTA) branch; the ROA; the RIPA, with a right vaginal artery (RVA) branch and a possible RMRA branch; and the RIGA.

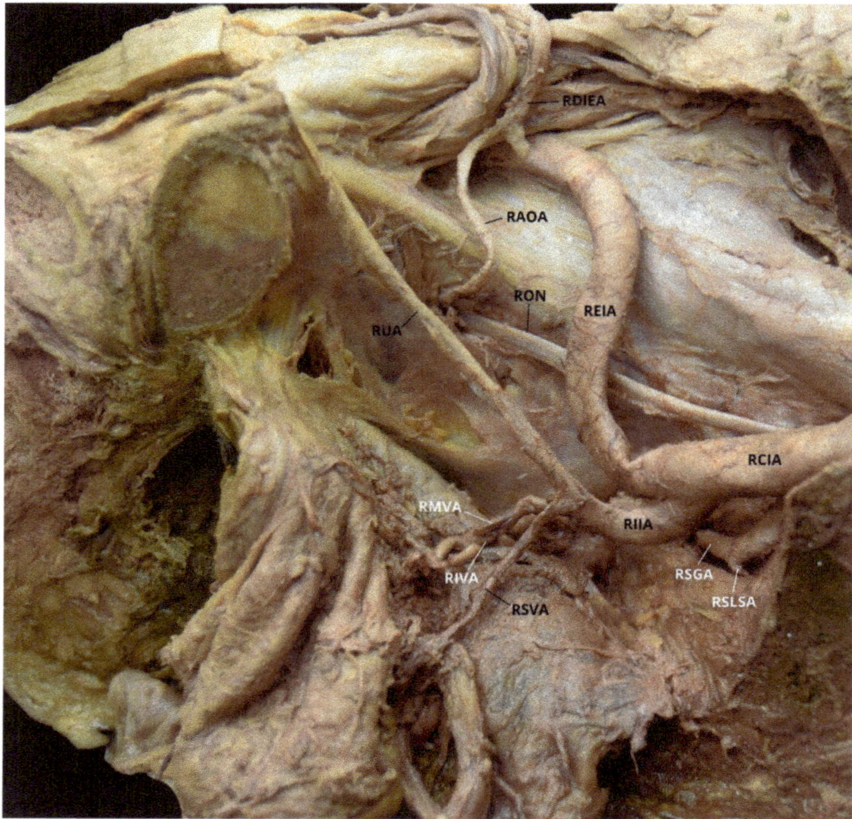

Figure 4. Facilitated display highlighting the right internal iliac artery branches with the right aberrant obturator artery for Case 2. RAOA = Right Aberrant Obturator Artery; RCIA = Right Common Iliac Artery; RDIEA = Right Deep Inferior Epigastric Artery; REIA = Right External Iliac Artery; RIIA = Right Internal Iliac Artery; RIVA = Right Inferior Vesical Artery; RON = Right Obturator Nerve; RMVA = Right Middle Vesical Artery; RSGA = Right Superior Gluteal Artery; RSLSA = Right Superior Lateral Sacral Artery; RSVA = Right Superior Vesical Artery; RUA = Right Umbilical Artery. The Right Inferior Gluteal Artery (RIGA), Right Iliolumbar Artery (RILA), Right Inferior Lateral Sacral Artery (RILSA), Right Internal Pudendal Artery (RIPA), and the Right Middle Rectal Artery (RMRA) are not pictured.

Case 4: A 91-year-old White Female with a listed cause of death of obstructive pulmonary disease was found to have a unilateral AOA on the left pelvic side with a Yamaki et al. (1998) Group B classification bilaterally and a Sañudo et al. (2011) Type B classification (Tables 1 and 2) [8,26]. The LIIA PDB included the LILA; the LSLSA; the LILSA; and the LSGA, with a LIGA branch. The LIIA ADB were the LUA, with a LSVA branch and a possible LUTA branch; and the LIPA, with a possible LMRA branch. There was significant damage from student dissection resulting in the inability to discern positively all of the branches present, and no LVA was found. The LAOA originated from the LDIEA (Figure 6). The RIIA PDB were the RILA; the ROA, with several arterial branches to the iliacus and iliopsoas muscles; the RSLSA; the RILSA; and the RSGA. The RIIA ADB included the RUA, with a RSVA branch and a possible RUTA; the RIGA; and a common branch of RIPA possibly with a RMRA. Due to significant damage from student dissection, the RVA was not found and several branches could not be named with certainty (Figure 7).

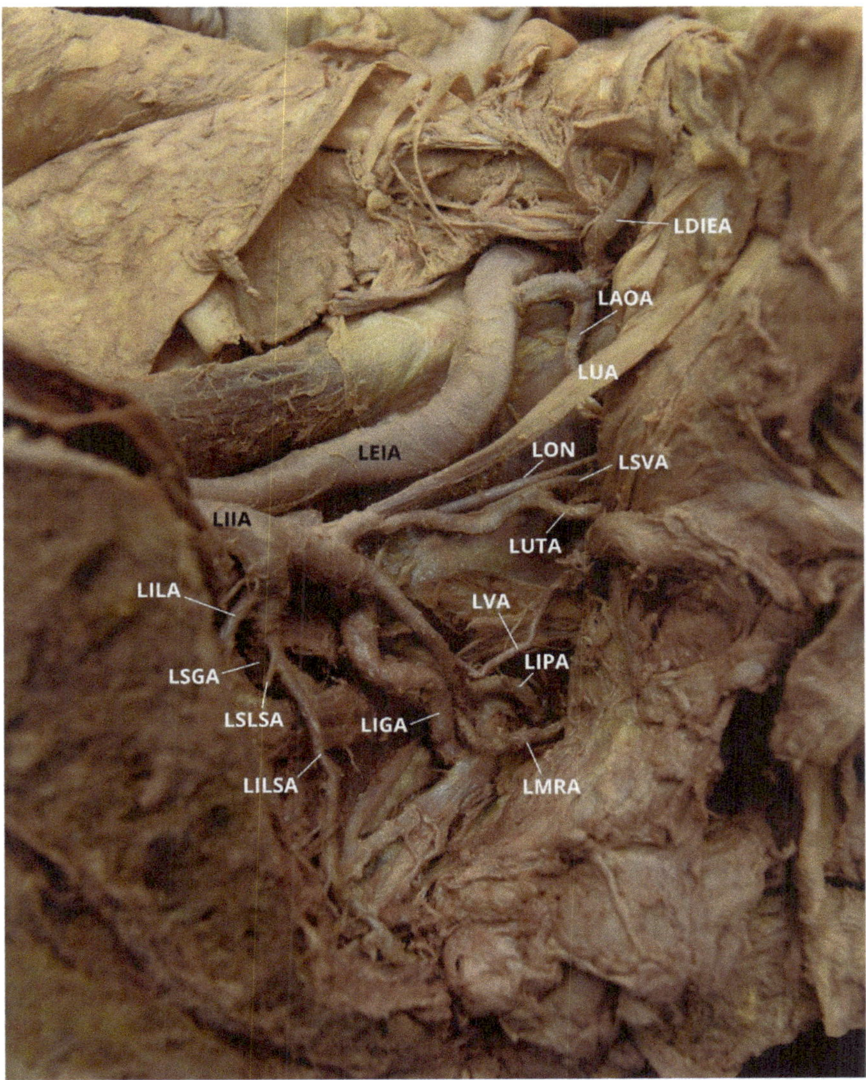

Figure 5. Facilitated display highlighting the left internal iliac artery branches with the left aberrant obturator artery for Case 3. LAOA = Left Aberrant Obturator Artery; LDIEA = Left Deep Inferior Epigastric Artery; LEIA = Left External Iliac Artery; LIGA = Left Inferior Gluteal Artery; LIIA = Left Internal Iliac Artery; LILA = Left Iliolumbar Artery; LILSA = Left Inferior Lateral Sacral Artery; LIPA = Left Internal Pudendal Artery; LMRA = Left Middle Rectal Artery; LON = Left Obturator Nerve; LSGA = Left Superior Gluteal Artery; LSLSA = Left Superior Lateral Sacral Artery; LSVA = Left Superior Vesical Artery; LUA = Left Umbilical Artery; LUTA = Left Uterine Artery; LVA = Left Vaginal Artery.

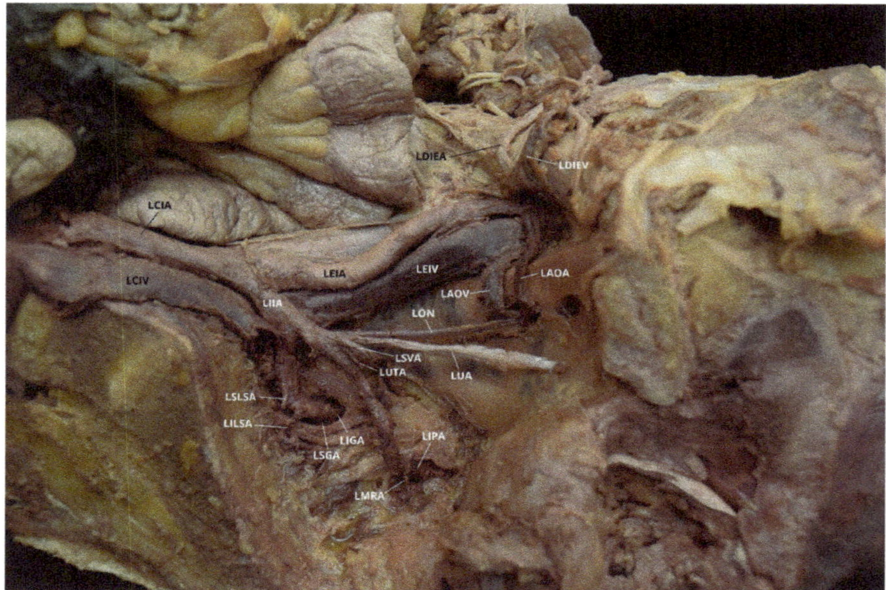

Figure 6. Facilitated display highlighting the left internal iliac artery branches with the left aberrant obturator artery for Case 4. LAOA = Left Aberrant Obturator Artery; LAOV = Left Aberrant Obturator Vein; LCIA = Left Common Iliac Artery; LCIV = Left Common Iliac Vein; LDIEA = Left Deep Inferior Epigastric Artery; LDIEV = Left Deep Inferior Epigastric Vein; LEIA = Left External Iliac Artery; LEIV = Left External Iliac Vein; LIGA = Left Inferior Gluteal Artery; LIIA = Left Internal Iliac Artery; LILSA = Left Inferior Lateral Sacral Artery; LIPA = Left Internal Pudendal Artery; LMRA = Left Middle Rectal Artery; LON = Left Obturator Nerve; LSGA = Left Superior Gluteal Artery; LSLSA = Left Superior Lateral Sacral Artery; LSVA = Left Superior Vesical Artery; LUA = Left Umbilical Artery; LUTA = Left Uterine Artery. The Left Iliolumbar Artery (LILA) and the Left Vaginal Artery are not pictured.

Case 5: An 84-year-old White Female with a listed cause of death of acute hemorrhagic cerebrovascular accident (CVA) exhibited bilateral AOAs with a Yamaki et al. (1998) Group A classification on the left, Group B on the right, and a Sañudo et al. (2011) Type B classification bilaterally (Tables 1 and 2) [8,26]. The LIIA PDB included the LILA, the LSLSA, the LILSA, and the LSGA. The LIIA ADB were the LUA, a common branch of LSVA and LUTA, the LVA, a common branch of LMRA and LIPA, and the LIGA. The LAOA originated from the LDIEA. A branch originating from the LAOA supplied the left obturator internus muscle. The LIIA PDB had a LAOAB providing multiple arterial branches to the left iliacus and iliopsoas muscles (Figure 8). The RIIA PDB included the RILA, a common branch of the RSLSA and the RILSA, the RSGA, and the RIGA. The RIIA ADB were the RUA, with a RSVA branch; the RUTA, with an RVA branch; and a common branch of the RMRA and the RIPA. There was also a right accessory obturator artery branch (RAOAB) from the RMRA that supplied the inferior aspect of the right obturator internus muscle. The RAOA originated from the RDIEA. There were two small branches originating from the RAOA that supply the superior aspect of the right obturator internus muscle (Figure 9).

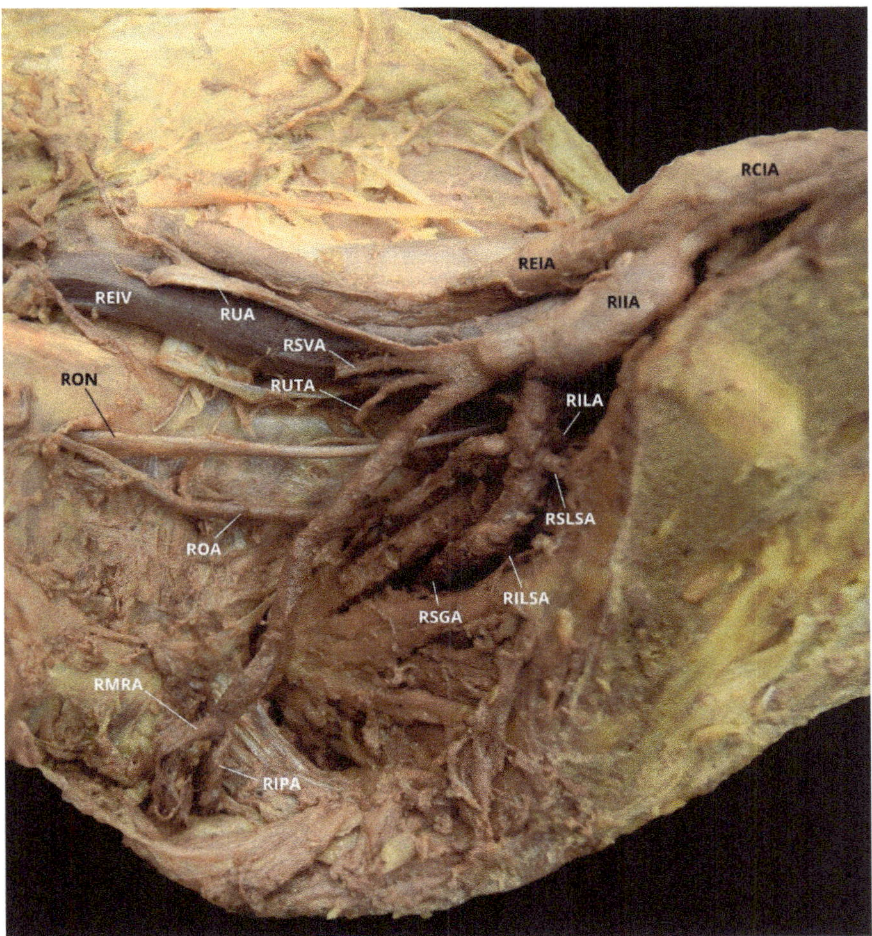

Figure 7. Facilitated display highlighting the right internal iliac artery branches for Case 4. RCIA = Right Common Iliac Artery; REIA = Right External Iliac Artery; REIV = Right External Iliac Vein; RIIA = Right Internal Iliac Artery; RILA = Right Iliolumbar Artery; RILSA = Right Inferior Lateral Sacral Artery; RIPA = Right Internal Pudendal Artery; RMRA = Right Middle Rectal Artery; ROA = Right Obturator Artery; RON = Right Obturator Nerve; RSGA = Right Superior Gluteal Artery; RSLSA = Right Superior Lateral Sacral Artery; RSVA = Right Superior Vesical Artery; RUA = Right Umbilical Artery; RUTA = Right Uterine Artery. The Right Inferior Gluteal Artery (RIGA) and Right Vaginal Artery (RVA) are not pictured.

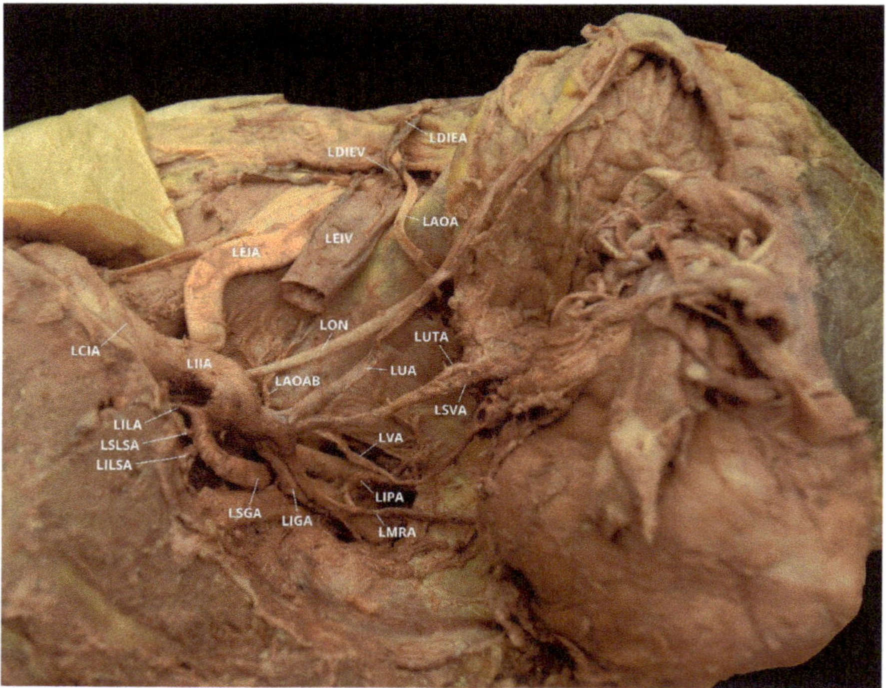

Figure 8. Facilitated display highlighting the left internal iliac artery branches with the left aberrant obturator artery for Case 5. LAOA = Left Aberrant Obturator Artery; LAOAB = Left Accessory Obturator Artery Branch; LCIA = Left Common Iliac Artery; LDIEA = Left Deep Inferior Epigastric Artery; LDIEV = Left Deep Inferior Epigastric Vein; LEIA = Left External Iliac Artery; LEIV = Left External Iliac Vein; LIGA = Left Inferior Gluteal Artery; LIIA = Left Internal Iliac Artery; LILA = Left Iliolumbar Artery; LILSA = Left Inferior Lateral Sacral Artery; LIPA = Left Internal Pudendal Artery; LMRA = Left Middle Rectal Artery; LSVA = Left Superior Vesical Artery; LON = Left Obturator Nerve; LSGA = Left Superior Gluteal Artery; LSLSA = Left Superior Lateral Sacral Artery; LUA = Left Umbilical Artery; LUTA = Left Uterine Artery; LVA = Left Vaginal Artery.

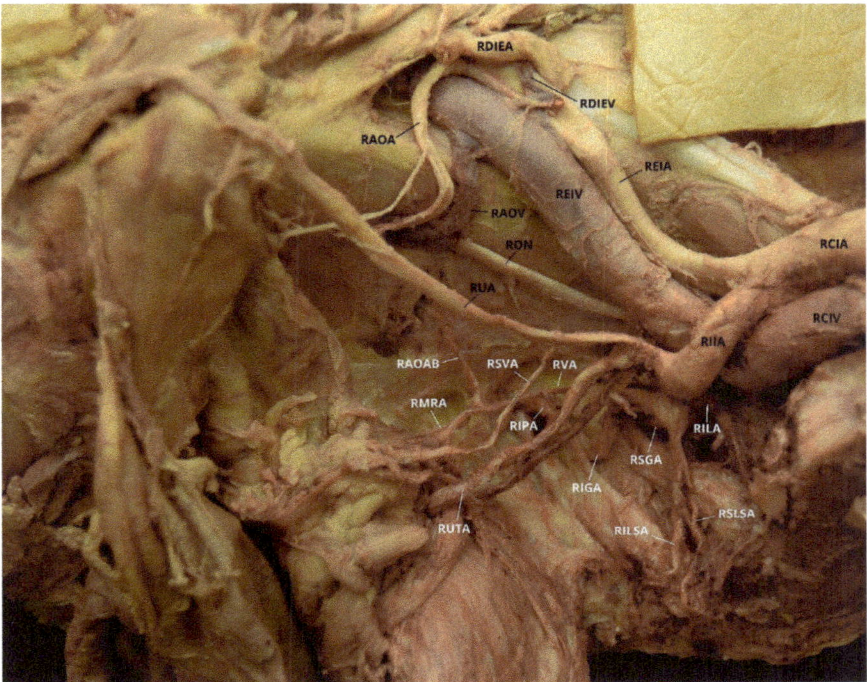

Figure 9. Facilitated display highlighting the right internal iliac artery branches with the right aberrant obturator artery for Case 5. RAOA = Right Aberrant Obturator Artery; RAOAB = Right Accessory Obturator Artery Branch; RCIA = Right Common Iliac Artery; RCIV = Right Common Iliac Vein; RDIEA = Right Deep Inferior Epigastric Artery; RDIEV = Right Deep Inferior Epigastric Vein; REIA = Right External Iliac Artery; REIV = Right External Iliac Vein; RIGA = Right Inferior Gluteal Artery; RIIA = Right Internal Iliac Artery; RILA = Right Iliolumbar Artery; RILSA = Right Inferior Lateral Sacral Artery; RIPA = Right Internal Pudendal Artery; RMRA = Right Middle Rectal Artery; ROA = Right Obturator Artery; RON = Right Obturator Nerve; RSGA = Right Superior Gluteal Artery; RSLSA = Right Superior Lateral Sacral Artery; RSVA = Right Superior Vesical Artery; RUA = Right Umbilical Artery; RUTA = Right Uterine Artery; RVA = Right Vaginal Artery.

Case 6: A 36 year-old White Female with a listed cause of death of metastatic breast cancer exhibited a unilateral AOA on the right pelvic side with a Yamaki et al. (1998) Group A classification bilaterally and a Sañudo et al. (2011) Type E classification (Tables 1 and 2) [8,26]. The LIIA PDB were a common branch of the LILA and the LSLSA, the LILSA, and the LSGA. The LIIA ADB included the LUA; the LSVA; the LUTA; the LOA; an accessory LILSA (ALILSA); a common trunk of the LMRA and the LVA; and the LIGA, with a LIPA branch. The LOA provided multiple branches supplying the obturator internus, iliopsoas, and iliacus muscles. There was no distinct separation between the anterior and posterior division on the left pelvic side (Figure 10). The RIIA PDB were the RILA, the RSLSA, the RILSA, and the RSGA. The RIIA ADB included the RUA, with a RSVA branch; the RUTA; a common trunk of the RVA and the RMRA; and the RIGA, with a RIPA branch. The RAOA did not originate from the RDIEA, but rather, as an independent branch from the REIA (Figure 11).

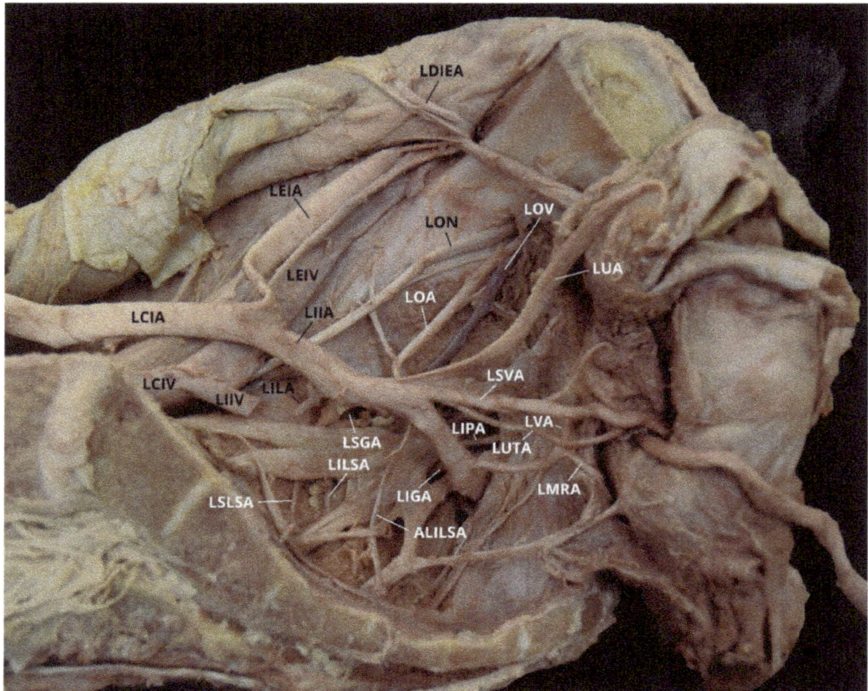

Figure 10. Facilitated display highlighting the left internal iliac artery branches for Case 6. ALILSA = Accessory Left Inferior Lateral Sacral Artery; LCIA = Left Common Iliac Artery; LCIV = Left Common Iliac Vein; LDIEA = Left Deep Inferior Epigastric Artery; LEIA = Left External Iliac Artery; LEIV = Left External Iliac Vein; LIIA = Left Internal Iliac Artery; LIGA = Left Inferior Gluteal Artery; LIIV = Left Internal Iliac Vein; LILA = Left Iliolumbar Artery; LILSA = Left Inferior Lateral Sacral Artery; LIPA = Left Internal Pudendal Artery; LMRA = Left Middle Rectal Artery; LOA = Left Obturator Artery; LON = Left Obturator Nerve; LOV = Left Obturator Vein; LSGA = Left Superior Gluteal Artery; LSLSA = Left Superior Lateral Sacral Artery; LSVA = Left Superior Vesical Artery; LUA = Left Umbilical Artery; LUTA = Left Uterine Artery; LVA = Left Vaginal Artery.

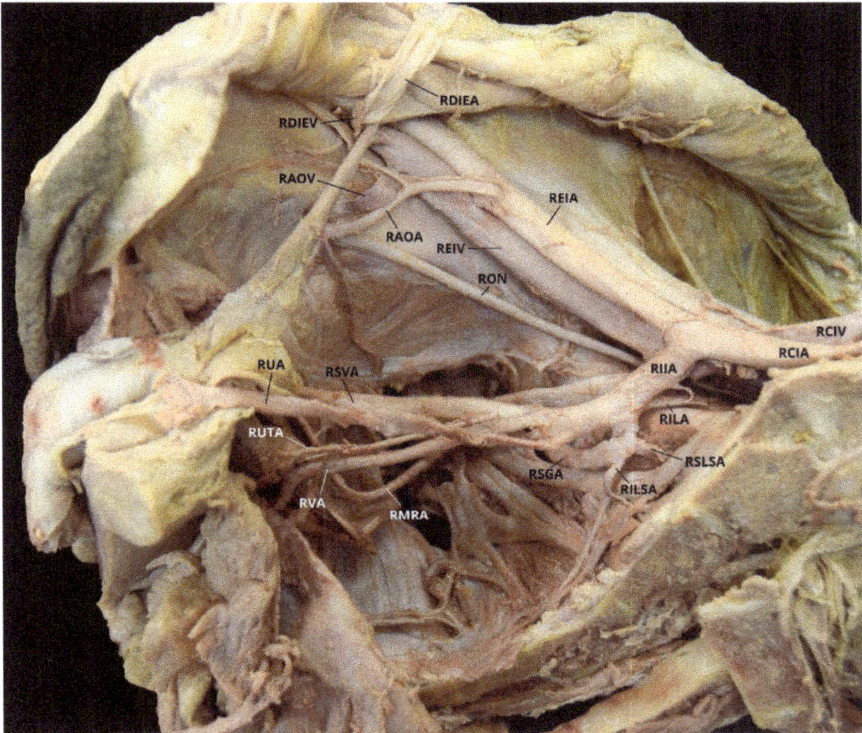

Figure 11. Facilitated display highlighting the right internal iliac artery branches with the right aberrant obturator artery for Case 6. RAOA = Right Aberrant Obturator Artery; RAOV = Right Aberrant Obturator Vein; RCIA = Right Common Iliac Artery; RCIV = Right Common Iliac Vein; RDIEA = Right Deep Inferior Epigastric Artery; RDIEV = Right Deep Inferior Epigastric Vein; REIA = Right External Iliac Artery; REIV = Right External Iliac Vein; RIIA = Right Internal Iliac Artery; RILA = Right Iliolumbar Artery; RILSA = Right Inferior Lateral Sacral Artery; RMRA = Right Middle Rectal Artery; RON = Right Obturator Nerve; RSGA = Right Superior Gluteal Artery; RSLSA = Right Superior Lateral Sacral Artery; RSVA = Right Superior Vesical Artery; RUA = Right Umbilical Artery; RUTA = Right Uterine Artery; RVA = Right Vaginal Artery. The Right Inferior Gluteal Artery (RIGA) and the Right Internal Pudendal Artery (RIPA) are not pictured.

Case 7: A 63-year-old White Male with a listed cause of death of renal cancer was found to have a unilateral AOA on the right pelvic side with a Yamaki et al. (1998) Group A classification bilaterally and a Sañudo et al. (2011) Type B classification bilaterally (Tables 1 and 2) [8,26]. The LIIA PDB included the LILA, the LSLSA, the LILSA, and the LSGA. The LIIA ASB were the LUA, with a branch of the LSVA; the LIVA; and a common trunk of the LOA, the LIPA, and the LIGA. The LIPA also had a common trunk with the LMRA. The RIIA PDB included the RILA, the RLSA, and the RSGA. The RIIA ASB were the RUA, with a RSVA branch; the RMVA; the RIVA; the RIPA, with a RMRA branch; and the RIGA. The RAOA originated from the RDIEA (Figure 12).

Figure 12. Facilitated display highlighting the right internal iliac artery branches with the right aberrant obturator artery for Case 7. RAOA = Right Aberrant Obturator Artery; RDIEA = Right Deep Inferior Epigastric Artery; REIA = Right External Iliac Artery; RIGA = Right Inferior Gluteal Artery; RIIA = Right Internal Iliac Artery; RIPA = Right Internal Pudendal Artery; RIVA = Right Inferior Vesical Artery; RMRA = Right Middle Rectal Artery; RMVA = Right Middle Vesical Artery; RON = Right Obturator Nerve; RSVA = Right Superior Vesical Artery; RUA = Right Umbilical Artery. The Right Iliolumbar Artery (RILA), Right Inferior Lateral Sacral Artery (RILSA), Right Superior Gluteal Artery (RSGA), and the Right Superior Lateral Sacral Artery (RSLSA) are not pictured.

3. Discussion

3.1. Obturator Artery

The abdominal aorta divides into the right and left common iliac arteries in the range of L3–L5, with the most common site being anterolateral to the left side of L4. Each CIA normally bifurcates into the EIA and IIA within the range of the L4–L5 disc and the mid-height of the S2 vertebra [11,13,27,28]. The EIA mainly supplies the lower limbs. The IIA usually descends posteriorly to the superior margin of the greater sciatic foramen where it divides into the posterior and anterior divisions. The posterior division passes back to the greater sciatic foramen, while the anterior division descends towards the ischial spine. These divisions provide many of the branches that supply the pelvic viscera, pelvic walls, perineum, and the gluteal region. These visceral and parietal arteries branch in numerous ways, and variations are common [1,11,23,26,28]. In the male, there are normally four visceral branches: the superior vesical artery (SVA), inferior vesical artery (IVA), middle rectal (or hemorrhoidal) artery (MRA), and the IPA. There is also a smaller branch called the prostatic artery, which is usually a branch of the IVA. In the female, there are normally five visceral branches: the SVA, uterine artery (UTA), vaginal artery (VA), MRA, and the IPA. In both males and females, there are normally six parietal branches: the ILA, superior lateral sacral artery (SLSA), inferior lateral sacral artery (ILSA), superior gluteal artery (SGA), IGA, and the OA. After birth, the umbilical artery (UA) becomes a ligament. The middle vesical (or vesiculodeferential) artery (MVA) usually derives from or is an adjacent branch to the IVA in males and the UTA or VA in females.

Yamaki et al. (1998) created a classification system (Groups A-D) for IIA branching patterns that was adapted from the Adachi (1928) classification method (Table 1) [1,26]. With the Yamaki et al. method, IIA branching patterns are classified based on the following three main branches: the SGA, IGA, and the IPA. The obturator artery is not used in the classification because of its high rate of origin variability from the IIA and EIA (Table 1). Group A is considered the basic IIA branching pattern because, of the identified cases, it occurs most frequently (60–80%). Group B is the second most frequent IIA branching pattern (15–30% of the population). Group C has been found in 5–7% of pelvic sides. Group D, a very rare pattern classification, has only been identified in one pelvic side (0.2%) [1,26].

The OA is a parietal extrapelvic branch of the IIA that usually arises from its anterior division (21–88.9% occurrence rate) with the obturator vein (OV) draining into the internal iliac vein (IIV) [3,5,7–14,21,23,27–32,34–39]. The OA usually arises either on the lateral or dorsolateral surface of the anterior division. It can, however, vary in its origin (6.6–63.63%) and has the greatest frequency of variation among the IIA branches [2–11,13]. In both males and females, the OA can arise from the CIA, the IGA (2–9%), the IPA (2–3.8%), a common trunk for the IGA and the IPA (10%), the ILA (1–3.33%), the EIA (1.1–4%), a branch of the EIA (8–33.3%), or by a dual root from both the IIA and the EIA sources (6.5%) [2,3,5,7,9–11,14,21,23,27–33]. The OA may also originate from the posterior division of the IIA (0.5–14.5%), usually as a branch of the SGA (2–16.1%) [2,7,9–12,21,23,28,30–34,37,38,40]. Additionally, the OA can have varying origins on the left and right side of the same pelvis [2,29]. In the studied population, the OA arose from the anterior division of the IIA bilaterally in ten of the eighteen individuals (55.6%). The OA branched from the posterior division unilaterally in two cadavers (11.1%), one on the left pelvic side and one on the right (Figure 7). The individual with a LOA originating from the left IIA posterior division also had a ROA arising from the IIA anterior division. The individual with a ROA arising from the right IIA posterior division also exhibited a LAOA (Figure 6). The OA may also have two (dual) or three origins (1–25%), and it is possible to have an accessory OA (30–40%) [5,9,22,25,29,38,39,41]. Two cadavers exhibited instances of accessory OAs, with one individual exhibiting it bilaterally (Figures 2, 8 and 9).

Sañudo et al. (2011) classifies the OA variations into six different types (A-F) (Table 2). Types A and B are the most common (35.5% and 22.5%, respectively) [8,29]. Type E is the second rarest (1.7% of cases). Pick et al. (1942) and Leite et al. (2017) are the only two articles known by the authors to have referenced Type F, the rarest OA variation type (1.66%), with one case described in each article [7,8,32].

After originating from the IIA, the OA traverses the lateral wall of the pelvis, inferior to the brim and enters the obturator foramen near its superior edge. Its usual course is medial to the obturator fascia; lateral to the ureter, ductus deferens, and peritoneum; and inferior to the obturator nerve. As the OA passes through the pelvis, it can provide several muscular and visceral branches to the iliac, vesical, and pubic regions [3,9,10,14,21,22,27–29,32,34,37,38]. The muscular branches can supply iliacus, iliopsoas, and obturator internus muscles. The OA also has branches that supply the iliac fossa. Prior to leaving the pelvic cavity, the OA can provide a pubic branch as a collateral circulation with the EIA system via the DIEA. It then passes from the pelvic cavity through the obturator canal to the medial compartment of the thigh. When leaving the pelvic cavity through the obturator foramen, the OA divides into two branches, the anterior and posterior branches that anastomose with the internal circumflex artery. The anterior branch supplies obturator externus, pectineus, adductor longus, adductor brevis, adductor magnus, and gracilis muscles. It terminates by anastomosing with the posterior branch of the obturator and medial circumflex femoral arteries. The posterior branch provides blood supply to the semimembranosus, semitendinosus, long head of biceps femoris, and the adductor magnus muscles. It then anastomoses with the IGA [3,11,21,27–29,32,42].

3.2. Aberrant Obturator Artery

An AOA is an anatomical variation in which the OA, a standard branch of the anterior division of the IIA, arises instead from the EIA. Previous studies have reported a frequency of as many as 55.1% of individuals [3,4,21,27,28]. It can arise from the DIEA (2.6–44%); directly from the EIA (1.1–10%); or from the femoral artery (FA) (1.1–1.66%) with the OV draining into the deep inferior epigastric vein (DIEV), the external iliac vein (EIV), or into the femoral vein (FV) [2,3,5,7,8,12,13,16,19,21–23,25,27–29,31–34,36–38,43–51]. Many studies have found an AOA origin from the DIEA to be more common in females than males. It is rarely found bilaterally [3,4,27,35,37]. Pai et al. 2009, however, reported a higher incidence of AOA origin from the DIEA in males (47%) than in females (26%) [5]. The AOA is also more frequently observed on the left rather than the right side of the pelvis [3,7]. The AOA then passes anteromedially to the external iliac vein, encircles the internal end of the femoral canal, passes over the pectineal ligament and descends towards the obturator foramen [14,24]. In the studied population, six of the seven individuals with unilateral or bilateral AOAs originated from the DIEA (33.3%) and one individual's AOA originated directly from the EIA (5.6%) (Figure 11). Of the six individuals with an AOA originating from the DIEA, three were male and three were female. The individual that had an AOA originating directly from the EIA was female. Of the seven AOA cases, five had LAOAs (27.8%) and four had RAOAs (22.2%).

3.3. Embryonic Development

During the fourth week of fetal life, the right and left dorsal aortae fuse caudal to the tenth dorsal intersegmental artery to form the descending aorta. The UA is the specialized paired ventral segmental branch passing through the connecting stalk, on each side. During the fifth week, the proximal part of each UA anastomoses with fifth dorsal lumbar intersegmental artery to form a new stem. This stem forms the dorsal root of UA, while the original ventral root of the UA degenerates. The dorsal root of the UA gives rise to two arterial plexuses (the abdominal and the pelvic). The pelvic plexus persists as the CIA and gives off branches that become the EIA and the IIA [8,21,29,32,34,40,42].

It appears that the OA forms as a result of uneven growth of an anastomosis of the EIA and IIA. In addition, the OA arises comparatively late to supply the medial side of the thigh. Both the uneven anastomosis growth and its late appearance may explain the occurrence of OA origin and trajectory variations [5,8,9,11,21–23,25,28,29,34,39,40,52]. In relation to the uneven anastomosis growth theory, some OA variations, such as its origin from the IIA posterior division, can be explained as vascular channels for anastomosis that persist in the posterior IIA division and those predestined for the OA may have disappeared or degenerated in the anterior division [5,11,13,25,27,28,40,50,53]. The AOA originating from the DIEA may be due to the underdevelopment or obliteration of a normal obturator at its origin and an enlargement of an anastomosis between the pubic branches of DIEA and OA behind the pubic body [35,47]. A dual origin of the OA may be interpreted as the presence of two source channels for the blood flow, one from the IIA and the other from the DIEA [5,50].

It should also be noted that OA origin variations may also occur later in life due to pathological conditions. These may involve venoclusive or arterial thromboembolic phenomenon and trauma or surgery in the pelvic area [12].

3.4. Clinical Significance

Accidental hemorrhage is the leading cause of obstetrical mortality in the United States of America. It is also the leading cause of maternal deaths in the developing world. Thus, a thorough understanding of the IIA branching patterns and their possible vascular variations is essential for obstetric surgeons [10,11,34,54]. Such knowledge is also crucially important for general surgeons and interventional radiologists performing other types of pelvic procedures (i.e., hernia repairs or pelvic fractures), as well as for anatomists teaching pelvic vasculature [3,19,23,27,29,32,38,39,50,51,55–57].

A vascular variant of the AOA that can be encountered in pelvic procedures is known as the corona mortis (CMOR). CMOR, meaning "the crown of death", involves vascular communication(s) between the OA and EIA or DIEA vessels that is present in 8.22–84% of patients [1,5,10,12–14,18,19,22,24,29,32,38,39,41,46,50,51,56–59]. The wide variation in CMOR incidence suggests there are ethnic or regional differences [19]. This is an arterial branch variation that usually originates from the EIA, DIEA, or coexists with the OA and anastomoses with it, creating an arc around the internal end of the femoral canal above the superior pubic ramus [1,14,15,29,39,57,58]. It can be unilateral or bilateral, and there seems to be no significant difference in its incidence between males and females [46,60–64]. This vascular variant earns its name due to the significant risk of death raised by its injury, which can lead to substantial hemorrhage and difficult hemostasis [10,11,14–20,32]. During open or laparoscopic hernia surgery, unrecognized injury to this vessel can lead to significant hemorrhage into the extraperitoneal space between parietal peritoneum and transversalis fascia. The typical course of this anomalous vessel encircling the superior end of the femoral canal (thus "crown") puts it at risk in open or laparoscopic hernia surgery. This is important given the current practice of sending patients home on the day of surgery. Unobserved hemorrhage especially in the elderly with compromised coronary circulation can be life threatening. Kashyap et al. (2019) and Sanna et al. (2018) found that the venous CMOR is much more prevalent than arterial CMOR and the majority of cases involved small caliber vessels (<4 mm). Anatomical case reports have identified a wide variety in the pattern and number of arterial or venous CMORs, and most cases demonstrated dissimilarity between left and right pelvic sides [12,18,19,24,38,51,59].

CMOR poses a risk in surgical procedures involving the inferior part of the anterior abdominal wall because the associated vessels run above and behind the superior pubic ramus in a relatively vertical direction [14,15,24]. These retropubic vessels are of paramount importance for surgeons treating pelvic and acetabular trauma, totally extraperitoneal (TEP) inguinal hernioplasties (especially during mesh fixation onto Cooper's ligament), herniorraphies, transcatheter embolizations, muscle graft surgeries, lymphadenectomies, catheterizations, and during IIA aneurysm repairs [3,5,8–14,18,19,23,27–29,32–34,37,46,55,59].

Dissection near the superior pubic ramus and Bogros space during surgical intervention must be conducted cautiously with advanced knowledge of such vascular variations [5,9,11,12,16,18,22,23,50,51]. Prior to any pelvic procedure, preoperative angiographic analysis of bilateral internal and external iliac systems should be conducted to ensure adequate evaluation of a potential collateral supply [19,46]. The presence of an AOA is not exclusively a risk factor for surgical complications, but it could also be beneficial if the IIA and its collateral blood supply were to be ligated or obstructed. The AOA and its branches could be a source of collateral circulation, especially the branch to the femoral head [5,9,14,23,29,32,50]. Knowing the vascular pattern and awareness to assess for possible OA variations can decrease the risk of iatrogenic injury and may modify the surgical and procedural approaches to minimize the postsurgical complications [3,25,27–29,34,46,55,59].

4. Conclusions

OA origin variations, such as AOAs, are common in the literature and frequent in occurrence. Proficient knowledge of pelvic vascular anatomy is essential for performing embolizations, revascularization procedures, treating pelvic fractures, laparoscopic herniorrhaphies, and obstetrical procedures. Performing preoperative angiographic analysis to know the pelvic vascular pattern and having the awareness to assess for possible OA variations can decrease the risk of iatrogenic injury. It may also modify the surgical procedures to minimize the postsurgical complications. Such familiarity is equally important for anatomy instructors to convey such information to their students on the presence and frequency of such vascular variations.

Author Contributions: Conceptualization, G.G.; methodology, G.G. and K.M.; validation, G.G. and K.M.; formal analysis, G.G. and K.M.; investigation, G.G. and K.M.; resources, G.G. and K.M.; data curation, G.G. and K.M.; writing—original draft preparation, G.G.; writing—review and editing, G.G., K.M., and G.W.; visualization, G.G., K.M., and G.W.; supervision, G.G. and K.M.; project administration, G.G. and K.M. All authors have read and agreed to the published version of the manuscript.

Funding: This research received no external funding.

Acknowledgments: We would like to thank the family of our donors for their beneficent contribution. Without their generosity, this article would not have been possible.

Conflicts of Interest: The authors declare no conflicts of interest.

Disclaimer: The opinions or assertions contained herein are the private ones of the author/speaker and are not to be construed as official or reflecting the views of the Department of Defense, the Uniformed Services University of the Health Sciences or any other agency of the U.S. Government.

References

1. Bilhim, T.; Pereira, J.A.; Fernandes, L.; Tinto, H.R.; Pisco, J.M. Angiographic Anatomy of the Male Pelvic Arteries. *Am. J. Roentgenol.* **2014**, *203*, 373–382. [CrossRef]
2. Bergman, R.A.; Thompson, S.A.; Afifi, A.K.; Saadeh, F.A. *Compendium of Human Anatomic Variation: Catalog, Atlas and World Literature*; Urban and Schwarzenberg: Baltimore, MD, USA, 1988; p. 86.
3. Lee, E.Y.; Kim, J.Y.; Kim, H.N.; Sohn, H.-J.; Seo, J.H. Variant Origin of Obturator Artery: A Branch of Inferior Epigastric Artery from External Iliac Artery. *Korean J. Phys. Anthropol.* **2013**, *26*, 125–130. [CrossRef]
4. Requarth, J.A.; Miller, P.R. Aberrant obturator artery is a common arterial variant that may be a source of unidentified hemorrhage in pelvic fracture patients. *J. Trauma.* **2011**, *70*, 366–372. [CrossRef] [PubMed]
5. Pai, M.M.; Krishnamurthy, A.; Prabhu, L.V.; Pai, M.V.; Kumar, S.A.; Hadimani, G.A. Variability in the origin of the obturator artery. *Clin. Sao Paulo* **2009**, *64*, 897–901. [CrossRef]
6. Kawai, K.; Honma, S.; Koizumi, M.; Kodama, K. Inferior epigastric artery arising from the obturator artery as a terminal branch of the internal iliac artery and consideration of its rare occurrence. *Ann. Anat.* **2008**, *190*, 541–548. [CrossRef]
7. Pick, J.W.; Barry, J.; Anson, B.J.; Ashley, F.L. The origin of the obturator artery: Study of 640 body halves. *Am. J. Anat.* **1942**, *70*, 317–343. [CrossRef]
8. Sañudo, J.; Mirapeix, R.; Rodriguez-Niedenführ, M.; Maranillo, E.; Parkin, I.; Vázquez, T. Obturator artery revisited. *Int. Urogynecol. J.* **2011**, *22*, 1313–1318. [CrossRef] [PubMed]
9. Mamatha, H.; D'Souza, A.S.; Ankolekar, V.; Bangera, H. An anomalous origin of obturator artery: A case report. *Indian J. Basic Appl. Med. Res.* **2014**, *3*, 293–295.
10. Sakthivel; Priyadarshini, S. Variability of origin of obturator artery and its clinical significance. *Int. J. Anat. Res.* **2015**, *3*, 1704–1709. [CrossRef]
11. Sonje, P.D.; Vatsalaswamy, P. Study of Variations in the Origin of Obturator Artery. *Indian J. Vasc. Endovasc. Surg.* **2016**, 131–135. [CrossRef]
12. Suryavanshi, S.; Bhagel, P.; Sharma, D. Variability in the Origin of the Obturator Artery: A Descriptive Cross-Sectional Cadaveric Study. *IJSS J. Surg.* **2016**, *2*, 16–19.
13. Tunstall, R. Internal iliac arteries. In *Bergman's Comprehensive Encyclopedia of Human Anatomic Variations*, 1st ed.; Tubbs, S., Shoja, M.M., Loukas, M., Eds.; John & Wiley & Sons, Inc.: Hoboken, NJ, USA, 2016; pp. 694–721.
14. Cerda, A. Symmetric Bilateral Aberrant Obturator Artery. *A Case Rep.* **2016**, *34*, 1083–1086.
15. Baena, G.; Rojas, S.; Peña, E. Corona mortis: Anatomical and clinical relevance and occurrence in a sample of the Colombian population. *Int. J. Morphol.* **2015**, *33*, 130–136. [CrossRef]

16. Lau, H.; Lee, F. A prospective endoscopic study of retro pubic vascular anatomy in 121 patients undergoing endoscopic extra peritoneal inguinal hernio-plasty. *Surg. Endosc.* **2003**, *17*, 1376–1379. [CrossRef]
17. Daeubler, B.; Anderson, S.; Leunig, M.; Triller, J. Hemorrhage secondary to pelvic fracture: Coil embolization of an aberrant obturator artery. *J. Endovasc. Ther.* **2003**, *10*, 676–680. [CrossRef]
18. Darmanis, S.; Lewis, A.; Mansoor, A.; Bircher, M. Corona mortis: An anatomical study with clinical implications in approaches to the pelvis and acetabulum. *Clin. Anat.* **2007**, *20*, 433–439. [CrossRef]
19. Kashyap, S.; Diwan, Y.; Mahajan, S.; Diwan, D.; Lal, M.; Chauhan, R. The Majority of Corona Mortia Are Small Calibre Venous Blood Vessels: A Cadaveric Study of North Indians. *Hip Pelvis* **2019**, *31*, 40–47. [CrossRef]
20. Astarci, P.; Alexandrescu, V.; Hammer, F.; Elkhoury, G.; Noirhomme, P.; Rubay, J.; Poncelet, A.; Lacroix, V.; Glineur, D.; Verhelst, R. Late Presentation of Bleeding from a Traumatic Obturator Artery Aneurysm, Successfully Treated by Endovascular Means. *EJVES Extra* **2005**, *10*, 77–80. [CrossRef]
21. Biswas, S.; Bandopadhyay, M.; Adhikari, A.; Kundu, P.; Roy, R. Variation of origin of obturator artery in eastern indian population—A study. *J. Anat. Soc. India* **2010**, *59*, 168–172. [CrossRef]
22. Prabaharr, P.S. The incidence of aberrant obturator artery in South Indian population—A cadaveric study. *Univ. J. Pre. Paraclin. Sci.* **2019**, *5*, 1–3.
23. Rajive, A.V.; Pillay, M. A study of variations in the origin of obturator artery and its clinical significance. *J. Clin. Diagn. Res.* **2015**, *9*, AC12–AC15. [PubMed]
24. Smith, J.C.; Gregorius, J.C.; Breazeale, B.H.; Watkins, G.E. The Corona Mortis, a Frequent Vascular Susceptible to Blunt Pelvic Trauma: Identification at Routine Multidetector CT. *J. Vasc. Interv. Radiol.* **2009**, *20*, 455–460. [CrossRef] [PubMed]
25. Tantchev, L.S.; Gorchev, G.A.; Tomov, S.T.; Radionova, Z.V. Aberrant obturator vessels in minimally invasive pelvic lymph node dissection. *Gynecol. Surg.* **2013**, *10*, 273–278. [CrossRef]
26. Yamaki, K.; Saga, T.; Doi, Y.; Aida, K.; Yoshizuka, M. A statistical study of the branching of the human internal iliac artery. *Kurume Med. J.* **1998**, *45*, 333–340. [CrossRef] [PubMed]
27. Al-Talalwah, W.; Al-Hashim, Z.; Al Dorzi, S.; Al Hifzi, H.; Yasky, A.; Al Mousa, H.; Soames, R. The clinical significance of the obturator artery in origin variability. *Indian J. Sci. Res.* **2016**, *7*, 61–65.
28. Dehmukh, V.; Sign, S.; Sirohi, N.; Baruhee, D. Variation in the Obturator Vasculature During Routine Anatomy Dissection of a Cadaver. *Sultan Qaboos Univ. Med. J.* **2016**, *16*, e256–e258. [CrossRef]
29. Goke, K.; Pires, L.A.S.; Leite, T.F.O.; Chagas, C.A.A. Rare origin of the obturator artery from the external iliac artery with two obturator veins. *J. Vasc. Bras.* **2016**, *15*, 250–253. [CrossRef]
30. Parsons, F.G.; Keith, A. Sixth annual report of the Committee of Collective Investigation of the Anatomical Society of Great Britian and Ireland (1895–1896). *J. Anat. Physiol.* **1897**, *31*, 31–44.
31. Braithwaite, J.L. Variations in origin of parietal branches of internal iliac artery. *J. Anat.* **1952**, *86*, 423–430.
32. Leite, T.F.O.; Pires, L.A.S.; Goke, K.; Silva, J.G.; Chagas, C.A.A. Corona Mortis: Anatomical and surgical description on 60 cadaveric hemipelvises. *Rev. Col. Bras. Cir.* **2017**, *44*, 553–559. [CrossRef]
33. Rusu, M.C.; Ilie, A.C.; Brezean, I. Human anatomic variations: Common, external iliac, origin of the obturator, inferior epigastric and medial circumflex femoral arteries, and deep femoral artery course on the medial side of the femoral vessels. *Surg. Radiol. Anat.* **2017**, *39*, 1285–1288. [CrossRef] [PubMed]
34. Jusoh, A.R.; Abd Rahman, N.; Abd Latiff, A.; Othman, F.; Das, S.; Abd Ghafar, N.; Suhaimi, F.H.; Hussan, F.; Sulaiman, I.M. The anomalous origin and branches of the obturator artery with its clinical implications. *Rom. J. Morphol. Embryol.* **2010**, *51*, 163–166. [PubMed]
35. Buchanan, A.M. *Manual of Anatomy, Systematic and Practical, Including Embryology*; CV Mosby Company: St Louis, MO, USA, 1914; Volume 1, p. 436.
36. Naguib, N.; Nour-Eldin, N.; Hammerstingl, R.; Lehnert, T.; Floeter, J.; Zangos, S.; Vogl, T.J. Three dimensional reconstructed contrast-enhanced MR angiography for internal iliac artery branch visualization before uterine artery embolization. *J. Vasc. Interv. Radiol.* **2008**, *19*, 1569–1575. [CrossRef] [PubMed]
37. Deaver, J.B. *Surgical Anatomy: A Treatise on Human Anatomy in Its Application to the Practice of Medicine and Surgery*; P. Blakiston's Son & Co.: Philadelphia, PA, USA, 1903. [CrossRef]
38. Nayak, S.B.; Shetty, S.D.; Shetty, P.; Deepthinath, R.; George, B.M.; Mishra, S.; Sirasanagandla, S.R.; Jumar, N.; Abhinitha, P. Presence of abnormal obturator artery and an abnormal venous plexus at the anterolateral pelvic wall. *OA Case Rep.* **2014**, *3*, 49–50.

39. Sakthivelavan, S.; Sendiladibban, S.D.; Aristotle, S.; Sivanandan, A.V. Corona mortis—A case report with surgical implications distally. *Int. J. Anat. Var.* **2010**, *3*, 103–105.
40. Kumar, D.; Rath, G. Anomalous origin of obturator artery from the internal iliac artery. *Int. J. Morphol.* **2007**, *25*, 639–641. [CrossRef]
41. Berberoğlu, M.; Uz, A.; Ozmen, M.M.; Bozkurt, M.C.; Erkuran, C.; Taner, S.; Tekin, A.; Tekdemir, I. Corona mortis: An anatomic study in seven cadavers and an endoscopic study in 28 patients. *Surg. Endosc.* **2001**, *15*, 72–75.
42. Standring, S. *Gray's Anatomy: The Anatomical Basis of Clinical Practice*, 39th ed.; Elsevier: Amsterdam, The Netherlands, 2005; p. 1044.
43. Adachi, B. *Das Arteriensystem der Japaner*; Die Kaiserlich Japanische Universitat zu Kyoto: Kyoto, Japan, 1928.
44. Jastchinski, S. Die Abweighchungen der arteriaobturatoria. *Int. Mschr. Anat. Physiol.* **1891**, *8*, 111–127.
45. Mahato, N.K. Retro-pubic vascular anomalies: A study of abnormal obturator vessels. *Eur. J. Anat.* **2009**, *13*, 121–126.
46. Herskowitz, M.; Walsh, J.; Lilly, M.; McFarland, K. Importance of Both Internal and External Iliac Artery Interrogation in Pelvic Trauma as Evidenced by Hemorrhage from Bilateral Corona Mortis with Unilateral Aberrant Origin off the External Iliac Artery. *Case Rep. Radiol.* **2019**, *6734816*, 1–4. [CrossRef]
47. Quian, J.; Sharpey, W.; Thomson, A.; Cleland, J.G. *Quain's Elements of Anatomy*; Thomson, A., Schaefer, E.A., Thane, G.D., Eds.; James Walton: London, UK, 1867; Volume 1, p. 455.
48. Bilgiç, S.; Sahin, B. Rare arterial variation: A common trunk from the external iliac artery for the obturator, inferior epigastric and profunda femoris arteries. *Surg. Radiol. Anat.* **1997**, *19*, 45–47. [CrossRef] [PubMed]
49. Jakubowicz, M.; Czarniawska-Grzesinska, M. Variability in origin and topotraphy of the inferior epigastric and obturator arteries. *Folia Morphol.* **1996**, *2*, 121–126.
50. Ram, K.S.; Gupta, T.; Chawla, K.; Aggarwal, A.; Gupta, R.; Sahni, D. Corona Mortis—A Case Report. *J. Anat. Sci.* **2015**, *23*, 28–30.
51. Al-Talalwah, W. A new concept and classification of corona mortis and its clinical significance. *Chin. J. Traumatol.* **2017**, *19*, 251–254. [CrossRef]
52. Petrenko, V.M. Development of the obturator artery in human prenatal ontogenesis. *Morfologia* **2000**, *118*, 51–53. [PubMed]
53. Al-Talalwah, W.; Soames, R. Internal iliac artery classification and its clinical significance: Original Communication. *Rev. Argent. Anat. Clin.* **2014**, *6*, 63–71. [CrossRef]
54. Cunningham, F.G.; Leveno, K.J.; Bloom, S.L.; Hauth, J.C.; Gilstrap, L.C.; Wenstrom, K.D. *Williams Obstetrics*, 22nd ed.; McGraw–Hill Professional: New York, NY, USA, 2005; pp. 7–8.
55. Gilroy, A.M.; Hermey, D.C.; DiBenedetto, L.M.; Marks, S.C., Jr.; Page, D.W.; Lei, Q.F. Variability of the obturator vessels. *Clin. Anat.* **1997**, *10*, 328–332. [CrossRef]
56. Nigam, V.K.; Nigam, S. *Essentials of Abdominal Wall Hernias*; International Publishing House Pvt. Ltd.: New Delhi, India, 2008; p. 84.
57. Rusu, M.; Cergan, R.; Motoc, A.; Folescu, R.; Pop, E. Anatomical considerations on the corona mortis. *Surg. Radiol. Anat.* **2010**, *32*, 17–24. [CrossRef]
58. Ates, M.; Kinaci, E.; Kose, E.; Soyer, V.; Sarici, B.; Cuglan, S.; Korkmaz, F.; Dirican, A. Corona mortis: In vivo anatomical knowledge and the risk of injury in totally extraperitoneal inguinal hernia repair. *Hernia* **2016**, *20*, 659–665. [CrossRef]
59. Sanna, B.; Henry, B.M.; Vikse, J.; Skinningsrud, B.; Pękala, J.R.; Walocha, J.A.; Cirocchi, R.; Tomaszewski, K.A. The prevalence and morphology of the corona mortis (Crown of death): A *meta*-analysis with implications in abdominal wall and pelvic surgery. *Inj. Int. J. Care Inj.* **2018**, *49*, 302–308. [CrossRef]
60. Henning, P.; Brenner, B.; Brunner, K.; Zimmermann, H. Hemodynamic Instability Following an Avulsion of the Corona Mortis Artery Secondary to a Benign Pubic Ramus Fracture. *J. Trauma Acute Care Surg.* **2007**, *62*, E14–E17. [CrossRef] [PubMed]
61. Ten Broek, R.P.G.; Bezemer, J.; Timmer, F.A.; Mollen, R.M.H.G.; Boekhoudt, F.D. Massive haemorrhage following minimally displaced pubic ramus fractures. *Eur. J. Trauma Emerg. Surg.* **2014**, *40*, 323–330. [CrossRef] [PubMed]
62. Loffroy, R.; Yeguiayan, J.M.; Guiu, B.; Cercueil, J.; Krausé, D. Stable fracture of the pubic rami: A rare cause of life-threatening bleeding from the inferior epigastric artery managed with transcatheter embolization. *Can. J. Emerg. Med.* **2008**, *10*, 392–395. [CrossRef] [PubMed]

63. Wilson, C.J.; Edwards, R. Massive extraperitoneal hemorrhage after soft tissue trauma to the pubic branch of the inferior epigastric artery. *J. Trauma Acute Care Surg.* **2000**, *48*, 779–780. [CrossRef]
64. Okcu, G.; Erkan, S.; Yercan, H.S.; Ozic, U. The incidence and location of corona mortis: A study of 75 cadavers. *Acta Orthop. Scand.* **2004**, *75*, 53–55. [CrossRef]

Perspective

Exploring Anatomic Variants to Enhance Anatomy Teaching: Musculus Sternalis

Andrew J. Petto [1,*], David E. Zimmerman [2], Elizabeth K. Johnson [3], Lucas Gauthier [4], James T. Menor [5] and Nicholas Wohkittel [6]

1 Department of Biological Sciences and Integrative Health and Assessment Unit, Department of Kinesiology, University of Wisconsin, Milwaukee, WI 53201-0413, USA
2 School of Medicine, St George's University, University Centre, True Blue Bay, Grenada 262G 5G , West Indies; dzimmerm@sgu.edu
3 Department of Social Sciences, Southern New Hampshire University, Manchester, NH 03106, USA; e.johnson4@snhu.edu
4 Advanced Physical Therapy & Sports Medicine Clinic, Marinette, WI 54143, USA; lgauthi25@gmail.com
5 Milwaukee Public Schools, Milwaukee, WI 53208, USA; menorjt@milwaukee.k12.wi.us
6 Froedtert & The Medical College of Wisconsin, Wauwatosa, WI 53226, USA; nickwoahkittel@gmail.com
* Correspondence: ajpetto@uwm.edu

Received: 14 June 2020; Accepted: 16 July 2020; Published: 22 July 2020

Abstract: The opportunity to encounter and appreciate the range of human variation in anatomic structures—and its potential impact on related structures, function, and treatment—is one of the chief benefits of cadaveric dissection for students in clinical preprofessional programs. The dissection lab is also where students can examine unusual anatomic variants that may not be included in their textbooks, lab manuals, or other course materials. For students specializing in physical medicine, awareness and understanding of muscle variants has a practical relevance to their preparations for clinical practice. In a routine dissection of the superficial chest muscles, graduate students in a human gross anatomy class exposed a large, well-developed sternalis muscle. The exposure of this muscle generated many student questions about M sternalis: its prevalence and appearance, its function, its development, and its evolutionary roots. Students used an inquiry protocol to guide their searches through relevant literature to gather this information. Instructors developed a decision tree to assist students in their inquiries, both by helping them to make analytic inferences and by highlighting areas of interest needing further investigation. Answering these questions enriches the understanding and promotes "habits of mind" for exploring musculoskeletal anatomy beyond simple descriptions of function and structure.

Keywords: M sternalis; anatomic variation; learning strategies

---

1. Introduction

"No two bodies are exactly alike." We repeat this adage in every anatomy class; but students rarely observe significant anatomic variation directly, unless they have access either to advanced imaging or cadaveric dissection. For students in clinical preprofessional programs, the direct comparison of sizes, shapes, proportions, and relationships among anatomic structures has several potential benefits that can enhance the appreciation—and often the clinical relevance—of anatomic variation in their career preparation.

Among the variants sometimes revealed in anatomic dissection, students oten encounter features that are rare, unexpected, or just "not supposed to be there". Since there are literally thousands of such variants documented in the literature [1], it is not surprising that these should appear occasionally in routine anatomy instruction, such as in laboratory dissection sessions. When our students encounter these variants, there is a unique opportunity for instructors to help students see these structures as more than simple curiosities, and to place them instead into a broader context of influences that produce an anatomic form and that may have a practical relevance in clinical practice.

One major emphasis of our anatomy curriculum is to promote the understanding of anatomic structure and function as the result of the dynamic interplay of several biologic processes. Often, even in courses that focus on human movement, there is a tendency in textbooks and reference materials to treat the gross anatomy of the musculoskeletal system as invariant. Curricular presentations of musculoskeletal anatomy often involve little more than a summary of location, attachments, innervations, and typical actions. But, when students uncover unusual anatomic variations in the gross anatomy lab, there is an opportunity for instructors and students to use these examples to deepen the understanding and appreciation of how the musculoskeletal system is formed and attains its usual presentation.

In this paper, we demonstrate this approach, taking advantage of the opportunity presented by an unusual anatomic feature—a large, unilateral M sternalis—to illuminate the various biologic processes that contribute to the appearance and function of the musculoskeletal components of adult human anatomy. This report is focused on the application of a pedagogic approach to learning anatomy: a problem-based exploration that guides students through the functional, developmental, and phylogenetic influences on musculoskeletal form and function. The object is to help students develop "habits of mind" based on a process of disciplined inquiry that provides a framework for understanding anatomic features that can also be applied to help students interpret the significance of anomalous or unusual musculoskeletal variants in the context of how the processes that produce the usual anatomic features can also produce the unusual.

In our human gross anatomy course for doctoral students in physical therapy (KIN525: Human Gross Anatomy), one group of students carrying out routine dissection of the thorax exposed a well-developed and relatively large M sternalis, which they discovered was not included in the reference materials and dissection guides used in the course. This discovery prompted both a high level of excitement and an opportunity to use this excitement to promote a deeper understanding of adult musculoskeletal morphology.

When an unusual variant appears in the anatomy lab, the first questions from students are "What is it?" and "Where did this come from?" In this report, we will use this example of students' isolation of an unusually well-defined M sternalis as an example for how anatomic variation can be a gateway to deeper learning about the variables that affect the anatomic features that they will encounter professionally. Answering students' questions about the appearance of this muscle helped us to explore both the usual anatomic features that they encounter in the human body and the unusual variants that present themselves on occasion.

The first place students look for answers, of course, is in their anatomic atlases and dissection guide. There are a few additional resources that contain detailed observations about anatomic variations in human anatomy [1–4]. Bergman et al. [1,2] have one of the most comprehensive resources for anatomic variations in humans, illustrating an impressive wealth of these variants in multiple organs and systems. Platzer is somewhat more accessible in that its descriptions of variants are included in the same parts of the text as related structures in the musculoskeletal, nervous, and vascular systems [4]. Diogo and Abdala combine a detailed search of descriptions of anatomy literature for humans and their primate relatives supplemented with Diogo's own careful dissection of relevant specimens [5]. Diogo and Wood's comprehensive volume contains a description of M sternalis in humans and other apes [3].

The initial review of available literature on M sternalis suggested that the appearance of this muscle is a "rare" variant (though what constitutes "rare" is seldom quantified). Its typical prevalence is estimated at 3-5% of individuals [6,7], but other sources estimate higher rates of up to 20% based on a literature review [8]. It is of note that Jelev et al. estimated that prevalence rates in European populations averaged 4.7% but that prevalence was higher in African (10%) and Asian (up to 20%) populations [8]. A systematic survey in a Chinese population (estimates prevalence rates around 6% [9]). These latter studies suggest some variations in prevalence among regional geographic populations, and the lower rates in the earlier reports may result from the tendency to collect data fromsubjects of western European ancestry. Eisler confirms this with his survey of reports from Europe, where reported prevalence was under 10% and in the Far East, where prevalence was up to 15% [10].

In modern times, M sternalis may be encountered in thoracic surgery or in diagnostic imaging, such as mammography [11–13], which may account for a slightly higher reported prevalence for this muscle in females than in males. However, most reports of this anatomic variant arise from serendipitous findings, like the one in our teaching lab.

The earliest report in the western anatomic literature appears to be from a series of short observations on anatomic variations in Cabrolio [14]. Turner [6] credits Cochon-DuPuy [15] with the earliest attempt to describe M sternalis in relation to the other superficial anterior muscles—in this case in association with Rectus abdominis. Most 19th century sources point to the work of Albinus as the source for the first systematic and detailed description of this muscle [16]. Albinus [16] describes this variant as a rare example of nature's playfulness (or perhaps trickery or mockery): *"rarum naturae ludentis exemplum."* The earliest report in English appears to be M'Whinnie whose review suggested that M sternalis (which he referred to as Rectus sternalis) was commonly considered to be associated with either Rectus abdominis or M sternocleidomastoideus because of its location, topography, and most common attachments [17].

Turner appears to be the first to examine enough cadavers to estimate a prevalence (21 of 651, or about 3%), and he noted several different arrangements of the muscle [6]. He also reviewed the literature available to him at that time and reported that variants of this muscle were named in the literature as M sternalis, M presternalis, M rectus sternalis, M sternalis brutorum, or M thoracicus [6,18].

Most sources describe a strap-like muscle parallel and slightly lateral to the sternum and superficial to the pectoralis muscles. The muscle may be truly bilateral (with right and left muscles about equal in length and mass, and having similar or identical mirror-image attachments). However, Jelev et al. illustrate eight different arrangements with many variations based on the locations of the muscle bodies and their attachments [8]. Four variants are classified as Type I with attachments on the lower ribs only on one side of the chest, even if the clavicular attachments are bilateral. The four Type II variants all have attachments on the lower ribs on both sides of the chest, often with a crossing over to attach to the contralateral clavicle (Figure 1; Gruber [19]).

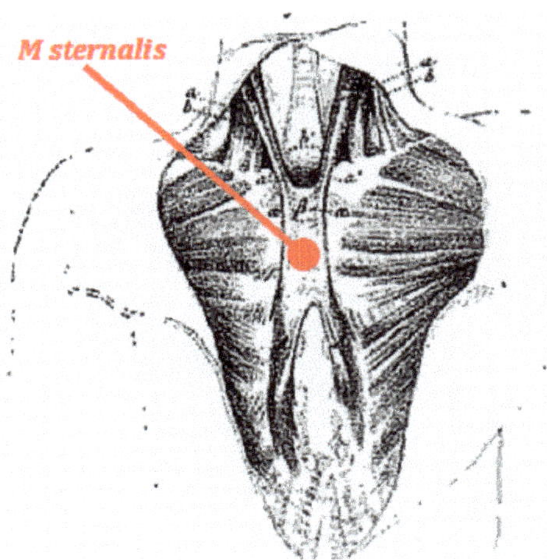

Figure 1. A bilateral M sternalis; the contralateral clavicular attachment is characteristic of Type II-1 in Jelev et al. [8]. Source: Figure 3 in Gruber [19]. Public domain based on publication date.

Although there seems to be general agreement on the location, attachments, and topographic associations of M sternalis in the literature, despite its anatomic variability, its innervation remains a matter of disagreement. But even case reports based on dissection of the muscle and its neurovascular supply have different findings.

Some report innervation only from intercostal nerves. Sarikçioğlu et al. identified the anterior cutaneous branch of intercostal nerve 6 as supplying the M sternalis muscle [20]. Natsis and Totlis and Hung et al. both identify the innervation as from a branch of intercostal nerve 2 [21,22]. Arráez-Aybar et al. report a "neurovascular pedicle" in the mid-portion of the M sternalis arising from the anterior cutaneous branch of intercostal nerve 3 (and supplied by anterior intercostal arteries from the internal thoracic artery and vein) [23].

Other sources identify innervation from the anterior or medial thoracic nerves. Katara et al. identify "twigs" of the pectoral nerve, but do not identify which [24]. Snosek et al. report that innervation is from the medial pectoral nerve [13]. Kida et al. are adamant that the only innervation is from the medial pectoral nerve [25]. And others, as far back as Eisler [10], suggest that the branches of the intercostal nerves that are seen in association with the M sternalis penetrate the muscle on their way to innervate the overlying skin or other nearby tissues, but do not serve the muscle itself [13,25,26].

To complicate things more, Pillay et al. report the innervation of a unilateral M sternalis via the medial pectoral nerve, but that of a bilateral M sternalis by intercostal nerves [27]. And Hung et al., also report finding innervation from both the intercostal and the medial pectoral nerves [22]. Among those reviewing the literature, both Vaithianathan et al. and Raikos et al. cite a report by O'Neil and Folan-Curran that 55% of cases indicate innervation by a pectoral nerve, 43% indicate innervation by intercostal nerves, and 2% indicate innervation by both [28–30].

Raikos suggests that the variability in innervation may have a developmental aspect [28]. The pattern of innervation may represent opportunistic connections between myocytes and neurons based on the topographic location of the precursor to M sternalis. There is also a significant number of reports of the appearance of M sternalis in anencephalic fetuses and infants [7,13,22,28,31]. However, the example we encountered occurs in an otherwise anatomically normal adult female, so major developmental anomalies appear to have no bearing on the appearance of this muscle in our case.

2. Initial Observation

Doctoral students in the physical therapy program at the University of Wisconsin-Milwaukee (UWM) uncovered a superficial band of muscle overlying the left lateral edge of the sternum and adjacent costochondral cartilages lying between the subcutaneous fascia and the pectoral muscles in an 87-year-old female during a routine dissection of the chest (Figure 2). A flat band of parallel fascicles was connected by a merger of the fascia cranially into connective tissues associated with the sternoclavicular joint. There was a similar merging between the fascia at the caudal extent of the muscle and the fascia associated with the M rectus abdominis. This muscle extended 18 cm from near the xiphoid process to the sternoclavicular joint and ranged from 1.5 to about 2 cm wide. There were no other unusual features uncovered in the dissection, including in the thorax, the remainder of the musculoskeletal dissection, or in the gross appearance of organs in the chest and abdomen.

The original dissection did not preserve all the contributing nerves or blood vessels, but the initial appearance was that the muscle was served by several nerves and vessels attaching at intervals along its length that emanated from the anterior chest wall. The impression was of nerves emerging from the intercostal spaces, but the individual nerves were not traced back to their spinal roots. The removal of overlying tissues before discovery of the M sternalis in this donor also made it impossible to verify the suggestions generated later by the literature review that these nerves might penetrate the muscle to serve the fascia and skin above the muscle layer.

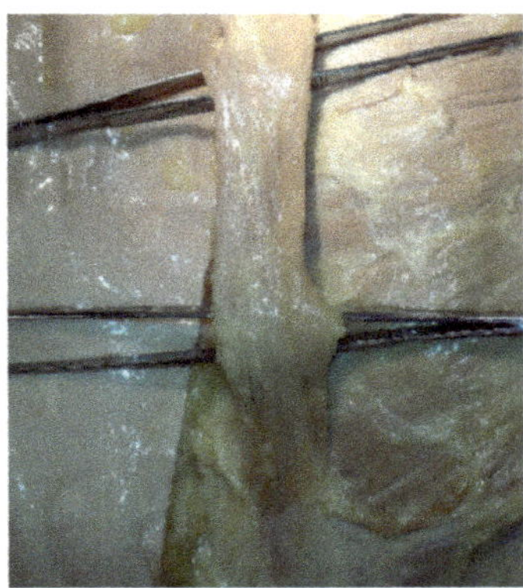

Figure 2. Photograph of M sternalis exposed during routine dissection in human gross anatomy laboratory; 25 cm forceps were used to indicate scale. Photograph by AJ Petto.

3. The Essential Questions

For any anatomic structure—perhaps especially for unusual variants—we want to learn why it appears as it does, if it has any function, and what its source is. We begin by exploring the typical explanations for such structural variation: Vestiges (or vestigial structures); anomalies (unusual development); remnants or rudimentary muscles (incomplete development via myogenesis or apoptosis), or atavisms (or "throwbacks"). These are described in more detail in Table 1.

Table 1. Explanations for skeletal muscle variants.

Label	Description	Example
Vestige	A rarely used muscle often appearing in a variable form, but with regular attachments, innervation, and vascularization. Any associated function is usually redundant with functions performed regularly by other muscles.	M palmaris longus or M plantaris; both are variably present and add little to the function of other muscles that share their attachments; their absence likewise has little or no impact on function.
Anomaly	A muscle that normally appears in humans, but its development is modified from its normal course.	Hypertrophy of M abductor digiti minimi or M quadratus plantae [7].
Rudimentary	A muscle that would normally appear in humans, but its development has been interrupted.	Embryonic muscles that fail to regress and lead to the persistence of a whole muscle or of small muscle slips as in M pectorodorsalis in individuals with trisomy 21 [7].
Atavism	A muscle that is normally present in our evolutionary ancestors and typically missing in humans, but which re-appears in humans.	Chondroepitrochlearis is located along the inferior surface of the Pectoralis major and inserts on the medial aspect of the intermuscular septum and medial epicondyle of the humerus. Diagnosed as a remnant of Panniculus carnosus [32] or as a derivative of the pectoralis group [33].

By exploring each of these potential explanations for the appearance of M sternalis in this donor, students can ultimately learn more about muscle function and development in general. Then they can apply this knowledge to muscle anatomy in other cases—for both the unusual and the expected anatomic presentation of skeletal muscles. In our labs, we use a mnemonic for engaging a deeper understanding of gross anatomic features: F•E•D•U•P (Function-Evolution-Development Understanding Protocol). This "protocol" guides student inquiry as they seek to learn more about specific aspects of any anatomic feature, but particularly in their exploration of information about unusual variants.

Function is the component of the mnemonic that is easiest for students to grasp. Function refers to the outcome of muscle action: when this muscle contracts, what are the effects on movement, position, or posture? Does the muscle's action cause or resist movement of any segment of the body? Does it result in or prevent a change in the relationship of the main axis of the body to the substrate (position)? Does it result in or resist a change in posture? These aspects of musculoskeletal function are typically available in most standard anatomy texts.

Evolution is the component of the mnemonic that most students have not considered—at least in the context of their studies in anatomy. Even those students who have experience in comparative anatomy—at a minimum, those who have dissected fetal pigs, cats, or other vertebrates in their anatomy or biology courses—tend not to think in evolutionary terms about the similarities and differences.

Evolution traces the peculiar combination of anatomic features that define the branching patterns in the tree of life. Branches are defined by the emergence of derived states of anatomic features that are shared by a group of descendants and their common ancestors but that separate them from organisms on other branches (for example, the lack of an external tail among all apes and their descendants; [34]). For students of human anatomy, the *phylogenetic* pattern illustrates the history of anatomic changes that the earliest humans inherited from populations of their ancestors and that are used as a basis for the anatomic specializations that define our species. Zanni and Opitz lay out a general approach to analyzing the evolutionary components of usual and unusual anatomic features [35].

Development is the third component, and, though students are aware of this aspect, they often have not given much thought specifically to how developmental processes result in the anatomic form they see in their dissections. Often our students are most interested in muscle conditioning and body building, and somewhat less in apoptosis and myogenesis, than in embryologic processes for muscle differentiation, migration, and attachment. Understanding how muscles come to be in their typical locations, with their typical sizes, shapes, attachments, innervation, and vascular supply is usually only challenged and brought to the forefront when a student uncovers a muscle with unexpected characteristics.

All these components together help us to resolve the essential questions and classify unexpected variants in the anatomy lab as indicated in Table 1. With the answers that result from asking about function, evolution, and development, students can use a graphic organizer—described in the discussion section—to apply the information they have uncovered to an anatomic variant. Students follow steps in a "decision tree" to a resolution of the essential questions or to indicate specific information based on direct observation of the dissection and from the literature review conducted in the process of applying the F•E•D•U•P protocol that students still need to locate before a step in the decision tree can be completed. The literature reported in the following sections reflects how students progress through the protocol. The goal is to locate relevant sources that address all three aspects of the protocol as they relate to observed anatomic features and to understand their contributions to the anatomic form they have observed.

3.1. Function

After answering the first question on exposing M sternalis—"What is it?"—the next question is often, "What does it do?" More specifically, when the muscle is activated and develops tension, is there any change in the position of skeletal elements located between its attachments? Students in our gross anatomy lab apply a standard template when studying all skeletal muscles to relate the function of a muscle with its location and attachments (See Table 2), and this can be applied to any muscle that presents in their dissections.

Table 2. Template for describing muscles.

Student Prompts	Example: Pectoralis Minor
Name this muscle	Pectoralis minor
Principal attachments	Anterior surfaces ribs 3–5; coracoid process of scapula
Muscle shape	Convergent as a whole, but individual slips can be parallel
Joint moved (for each joint that is located between the principal attachments)	Scapulothoracic and sternoclavicular
Functional characteristics of joint (plane, pivot, gliding, etc.)	Gliding (S-T) and sellar (S-C)
Type of movements allowed by joint	Depression/elevation; protraction/retraction; rotation
Planes of movements allowed by joint	Frontal and transverse
Directional relationship between proximal and distal attachments with respect to joint; for example, inferolateral to superomedial, etc.	Inferoanteromedial to superoposterolateral
Muscle's "Line of Pull" (orientation of main axis of muscle action relative to the segments connected by the joints) in the anterior–posterior and medial–lateral axes; for example, anterolateral or anteromedial, etc.	Inferoanteromedial
These features of the muscle combined with the movement allowed at the joints, causes this change in position	Depression, abduction, and medial rotation of scapula; downward rotation of glenoid fossa (also clavicular depression and protraction of shoulder girdle)
Of this segment of the body	Scapula, clavicle, shoulder girdle collectively
In this (these) plane(s)	Depression and rotation: frontal; protraction: transverse
Innervation	Pectoral nn (C6–C8)

Students in the Human Gross Anatomy course take their findings from this template to a concurrent course, Introduction to Physical Therapy Practice Examination Techniques (KIN526). In this concurrent course, they practice locating musculoskeletal structures by physical examination and consider the clinical implications of any atypical findings.

Generally speaking, it is easier to answer this question when the appearance of a muscle is regular and consistent from one individual to the next. However, the reports of the location, shape, and attachments of M sternalis indicate that there is considerable variability in its morphology [8,28].

We expect that a muscle that appears highly variable anatomically would be less likely to have any essential function. This is not to say that the muscle could never have any effect on the skeletal elements located between its attachments, but rather that such an effect might be idiosyncratic—dependent on the specific muscle morphology, and not generalizable in a way that would apply to all the ways that the muscle can appear. Any regular action that would affect the positions of skeletal elements engaged by the muscle body would probably rely on other, more regular muscles as their primary movers.

Until recently, examples of M sternalis have been reported almost entirely from studies of cadavers, so function could be inferred, but not confirmed. One exception comes from Kirk who was able to show the surface definition of the muscle under tension [36] (Figure 3). The author produced this effect "with the recti abdominis in flexion of the trunk, and, as is shown, with the pectorales in adduction of the arms; and its origin could then be seen to spread out transversely over the lower part of the Pectoralis major".

This description does not provide a clear answer on specific function, since the postural changes include both the adduction of the left arm by the Pectoralis major and the flexion of the trunk by the Rectus abdominis. However, it does give support to the two major candidates for the source of this muscle: P major and R abdominis, even if it does not resolve the question in favor of one or the other [13].

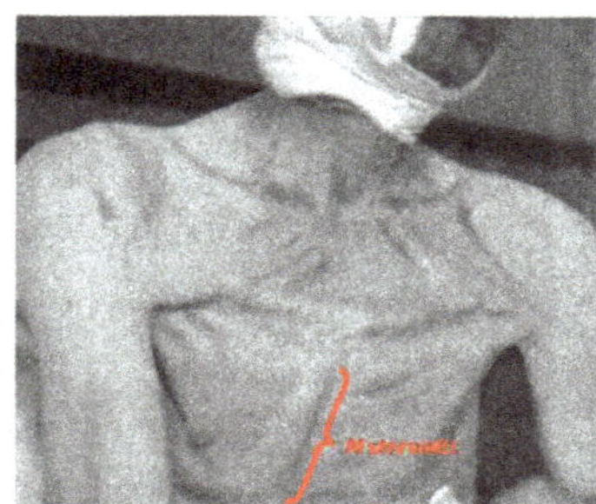

Figure 3. Image from Kirk showing tension in a unilateral, right M sternalis [36]. Image in the public domain based on publication date.

3.2. *Evolution*

The search for the evolutionary foundations of human morphology begins in comparative anatomy. Table 3 provides a summary of the search through the anatomic literature for the appearance of chest muscles in addition to the muscles of the pectoral girdle or the intercostals. In Urodeles, Hildebrand shows a continuous ventral muscle body from the pubis to the cephalad border of the sternum: a homolog of the R abdominis [37]. Omura et al. confirm that the Rectus group is active in maintaining posture on land in resistance to vertebral bending under gravity, thus their main function does not appear to involve ventilation of the lungs [38].

Table 3. Appearance of muscles similar to *M sternalis* in humans and other species.

Taxa	Description	Name	Appearance	Source
Urodeles	Lengthwise along ventral body wall between girdles	Rectus abdominis "group"	Not specified	[37,38]
Ruminants	Thin muscle from first rib to sternum and costal cartilages 3–5	Rectus thoracis	All species, though reduced in sheep and goats	[39]
Equines	Thin muscle from first rib to costal cartilage 4 and aponeurosis of Rectus abdominis	Rectus thoracis	Domestic horses	[39]
Hominids	Strap-like muscle from sternoclavicular joint or first rib variably to cranial attachment of Rectus abdominis	Sternalis	Variable appearance in *Homo* and *Hylobates syndactylus*	[3]
Baboons	Extension of Rectus abdominis aponeurosis to attach to manubrium	Rectus abdominis	Varies by species in genus *Papio*	[40]
	Aponeurosis with fleshy stratum	Rectus thoracis	*Papio papio*	[41]
Macaques	Aponeurosis fused with Rectus abdominis	Sternocostalis, Rectus thoracis, Rectus sternalis	*Macaca irus*, variably in *M sylvana*	[42]
	Extension of Rectus abdominis aponeurosis to attach to manubrium	Rectus abdominis	*Macaca mulatta*	[43]
Langurs and Tarsiers	Rectus abdominis extends to 1st costal cartilage	Rectus abdominis (abdominothoracic musculature)	*Semnopithecus*, occasionally in *Tarsius*	[44]
New World Monkeys	Rectus abdominis extends variably high into chest	Rectus abdominis	Various Hapalidae and Callimiconidae; genus *Cebus*	[45,46]

In mammals, Getty reports a Rectus thoracis muscle in ruminants and horses that lies over the ventral chest wall and extends from the cephalad aspect of the R abdominis to the top of the sternum [39]. In these animals, Getty reports that the muscle appears to assist in expansion of the chest cavity under conditions of aggressive inhalation, such as when running [39].

Among the primates, there are two patterns of ventral muscles worth noting. First, the R abdominis generally tends to attach more cephalad than is typical in humans [35,36], often on the manubrium [40,43,44] or as high as the first rib [44]. The naming of additional muscle bodies that are described as cephalad extensions of the aponeurosis of the R abdominis varies to include the names found in the older literature (discussed above). Osman Hill refers to this additional muscle as Rectus sternalis, while Diogo and Wood, only in hominids, refer to the muscle as simply M sternalis as found in *Homo* and *Hylobates syndactylus* (the siamang) [3,42].

Turner points out that part of the confusion about the attachments of R abdominis may derive from the anatomic work of Galen who used dissections of nonhuman primates as the basis for at least some of his descriptions of human anatomy [6]. Therefore, Turner argued, early anatomists were misled about the cephalad attachment of the R abdominis in normal human cadavers [6]. If this is the case, these anatomists might conclude that the appearance of an additional superficial strap-like ventral muscle extending to the sternum or clavicle might simply represent an elongation of or variation in the R abdominis.

Another candidate suggested as the basis for M sternalis is the Panniculus carnosus [6,47,48]. In other mammals, this muscle is typically located below the adipose layer lying deep to the dermis and above a connective tissue layer that separates the integument from the underlying skeletal muscle layer [48]; but it does not have any direct skeletal attachments [47]. Generally speaking, P carnosus does not appear in the higher primates (including hominins) as a distinct muscle of the trunk. However, Langworthy argues that this muscle is derived from the pectoral group in mammals that retain it, and cites Eisler's (1912) argument that M sternalis is formed by a failure of proper development in the pectoral musculature [10,47]. Bergman et al. also suggest that remnants of P carnosus may be found in the pars abdominalis of the pectoralis muscles or as "extra, independent, muscular slips from the abdominal aponeurosis which spreads forward on the rectus sheath" (https://www.anatomyatlases.org/AnatomicVariants/MuscularSystem/Text/P/05Panniculus.shtml) [1].

Naldaiz-Gastesi et al. identify 12 other muscles or muscular structures of the trunk and neck as potentially derived from, or containing, remnants of P carnosus, the most obvious of which is the Platysma [48]. These remnants, they argue, perform many functions—at least in nonhuman species—and some are incorporated into other regular, named muscles in the trunk. Based on their comparisons of the form and function of P carnosus in several mammalian species, Naldaiz-Gastesi et al. conclude that the innervation of P carnosus is independent of that of the underlying skeletal muscles allowing it to function separately from these other muscles [48]. If P carnosus is the source of M sternalis, then this difference in innervation might be consistent with the lack of apparent movement of skeletal elements by the M sternalis noted in the human anatomy literature.

However, these similarities do not by themselves help us to resolve the question of the evolutionary foundations of the appearance of M sternalis. Once we have described the patterns of similarities and differences in related organisms, we use these patterns to construct a cladogram or phylogenetic tree to establish the pattern of "descent with modification" that an evolutionary analysis requires [5]. In general, we are looking for an anatomic change that is established in an ancestral population and is shared by all the descendants of that population. It is also possible to define a group by a shared absence of a feature common in all its evolutionary relatives, because its ancestors have modified or eliminated it.

For example, a post-anal tail is considered a shared, conservative trait for the vertebrates, but none of the hominids (humans and other apes) retain this feature. The *loss* of the feature is a shared derived trait that helps to define the organisms on the ape evolutionary branch from those on other primate branches [34]. However, the loss of tail in some of the other primates or in other non-primate vertebrate species is understood as a repeatedly derived trait, that is, a feature that appears similar in several species but does not arise by common descent [49].

When we superimpose the patterns of ventral muscular anatomy on the primate cladogram (Figure 4), it is clear that M sternalis fails the test of a shared derived trait among the primates or even the hominids. When this muscle—or other muscles that may or may not be the same—appears in our phylogenetic diagram, it follows the pattern of a repeatedly derived trait: one that arises only on side branches unique to specific taxa, rather than at locations in the cladogram that join several lineages together (nodes) by virtue of their sharing this feature by way of descent from a common ancestor.

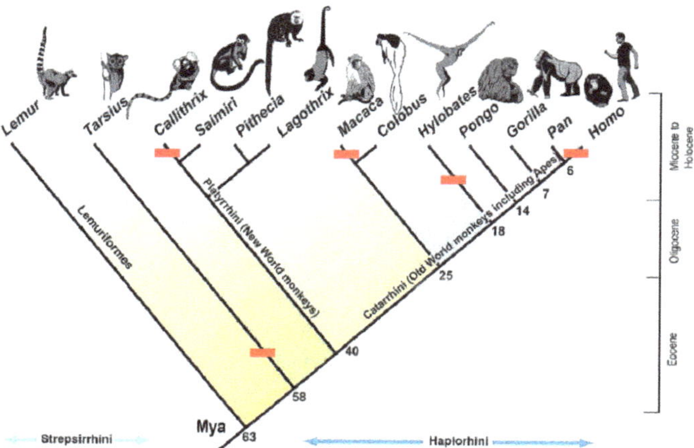

Figure 4. Cladogram showing the appearance and pattern of phylogenetic inheritance for muscles identified as potentially homologous with M sternalis in humans. Red bars indicate the branch on which the character state (presence of M sternalis or homologous muscle) appears. Data used for the character states described in Table 3. Image source: Available under Creative Commons Licensing from https://upload.wikimedia.org/wikipedia/en/thumb/f/f0/PrimateTree2.jpg/1280px-PrimateTree2.jpg.

3.3. *Development*

Our students do not typically study the details of the developmental processes that produce the muscles they encounter in the gross anatomy lab. For most of them, any background in developmental biology is limited to sections in their introductory biology or anatomy-and-physiology texts, or from a general overview, such as is found in Wolpert [50]. As a result, they are aware of the basics: muscles form from the dorsolateral aspects of the somites [51], transcription factors cause cellular differentiation, coalescence into muscle tissue is mediated by adhesion molecules (though it is possible for cells in these early tissues to dissociate and re-associate with other cells) and the final shape of the tissue can change in this process, *Hox* genes provide positional information, but developing muscles can be influenced by external conditions: the epigenetic influences from extracellular components and the availability of attachment sites in the underlying morphology [50]. Wolpert describes the ultimate musculoskeletal attachments as "democratic" and elsewhere as "promiscuous" in that these tissues and their associated connective tissues will attach to any appropriate substrate in their vicinity [50].

Understanding the contingent nature of much of developmental biology is perhaps the single greatest challenge for how students view developmental processes at this early stage of morphogenesis and their influences on adult anatomy. They need to develop the appreciation that development does not follow strict "blueprints" so much as a general schematic for the final form and location of skeletal muscles, and that the final result can be affected by numerous influences along the way.

Perhaps the most useful approach for students in the gross anatomy lab is to partition developmental processes into the effects whose results are more readily observable in the adult cadaver they encounter in the gross anatomy lab. These would look for evidence of (1) differentiation into muscle tissue from myocytes; (2) migration of muscle tissue to proper locations; and (3) formation of attachments appropriate for normal function.

3.3.1. Differentiation

Postcranial musculoskeletal development is remarkably conserved in vertebrates, such that processes relevant to human development can be observed in different model organisms [52]. Differentiation of the cells destined to be muscles are influenced by a muscle transcription factor (*MyoD*) to produce myoblasts [51]. Several myogenic regulatory factors are involved in the specification and differentiation of muscle tissues, and Pownall et al. detail multiple influences on myogenesis at different locations in the embryo [53].

According to Shearman and Burke, myoblasts are not committed a priori to specific muscles, but the connective tissues with which the myoblasts associate will dictate their final destinations [52]. These associations produce muscle bundles that retain their segmentation in the thoracic region, while the ventral portion of the bundles that will become the abdominal muscles fuse into the Rectus abdominis [51]. The presomitic mesoderm that will populate the thorax produces the connective tissue template for thoracic muscles regardless of where they are finally located [52]. Pownall et al. confirm that ectopic development of muscle masses in experimental studies is rare; that is, how a muscle develops depends a great deal on where its precursors are located [53].

Mekonen et al. indicate that the musculoskeletal primordia of the thorax appear as recognizable tissues by 5.5 weeks in the human embryo, and the establishment of the abdominal muscles is complete by week 10 [54]. For the M sternalis, a problem in this part of the process should be evident in malformation or other defects of the muscle tissue [54].

3.3.2. Migration

Typically, the development of bone, tendon, and muscle that will form a functional unit is coordinated to produce a functional whole [55]. Pownall et al. [53] report that there is a relatively small mass of migratory cells for populating the embryonic body wall and limbs, and that a myogenic factor (*MyoD*) is responsible for differentiation of muscle masses "in coordination with tendon and bone formation".

In several documented examples of problems in pectoral muscle development, underlying malformations in the connective tissues associated with skeletal attachments are common, for example in Poland syndrome (https://omim.org/entry/173800). In a case study of atypical muscle formation associated with Mm pectorales, Bannur et al. review atypical formations of pectoral musculature and suggest that the locations of muscle attachments can be useful in tracing their developmental histories in the interplay between migration, fusion, and apoptosis [56].

The atypical appearance of these muscle variants is often associated with at least one uncharacteristic attachment. This condition is consistent with Wolpert's characterization of muscle attachment as "democratic", in which skeletal muscles form attachments to connective tissue structures in nearby locations, rather than searching for fixed, pre-set attachment points [50]. For the M sternalis, a problem in this part of the process should be evident in atypical formations of connective tissues associated with the muscle.

3.3.3. Attachments

Early muscle development involves activation of myogenic factors, differentiation of myoblasts, and proliferation of myocytes. However, as Hasson reports, the final association between these muscle masses and specific skeletal muscles is not predetermined [57]. The role of the connective tissues within and surrounding the developing muscles helps to determine the patterning of these muscles that will result in the gross muscle anatomy we see in dissection.

Hasson (2011) reviews experimental studies on how the patterning in developing skeletal muscle arises and how the associations between the muscles and the related skeletal elements may be uncoupled. In our case of M sternalis, at least one attachment can be associated with those proposed as the ultimate source of the muscle (Rectus abdominis or Pectoralis major), so the possibility of a dissociation between one of the attachments and its intended skeletal target, as proposed in Bunnar et al., needs to be explored [56]. Freed from the initial association with a "typical" target, Wolpert's characterization of flexibility in acquiring final skeletal attachments must be considered [50].

For the M sternalis, a problem in this part of the process should be evident in relative consistency in the appearance of the muscle, but with variations in the points of attachment. However, at least one of the attachments should be consistent with the expected locations of regular, named muscles.

4. Discussion

In the gross anatomy lab, students occasionally encounter atypical arrangements of skeletal muscle. Since the donors in our labs are usually older adults, it is easiest to observe gross anatomic features—size, shape, location, attachments—which only go part of the way in helping them to understand the appearance of the variant and its relationship to other anatomic structures.

Although it is often described as a "rare" variant [6,7], the appearance of M sternalis was documented as early as the 17th century [14], and its identification as a known anatomic variant is common through the early 20th century [10,13,20–30], after which it tends to appear only chiefly in case reports of unusual anatomic features. This may be because M sternalis appears to be of little clinical significance except for those who might encounter it in medical imaging or surgical professions [11–13,28,29].

The consensus on the muscle's anatomic relations tends to shift, but current opinion seems mostly split between the Pectoralis major and the Rectus abdominis as potential sources for this muscle, based on its location, attachments, and innervation. The most disagreement seems to center on the innervation of this variant when it appears [10,13,20–30].

Phylogenetic analysis (Figure 4) shows that M sternalis is unlikely to be an "atavism" (see Table 1) failing most of the criteria proposed by Zanni and Opitz [35]. Since M sternalis usually appears in the absence of underlying skeletal or other connective-tissue abnormalities as it did in our lab, its presence is more likely due to local environmental influences on muscle development [6,20]. However, the wide variety of appearances of this variant does not indicate a single, consistent developmental driver of its presence, such as one might expect in known developmental disorders, such as Poland syndrome (https://omim.org/entry/173800) or various chromosomal mutations in which muscle development progresses atypically [8,58].

When confronted by such atypical skeletal muscles in gross anatomic dissection, there is an opportunity for enhancing learning with a deeper understanding of the structure and function of the human body. In the case of M sternalis, the first two essential questions—"What is it?" and "What is its function?"—are relatively easy to answer. The third—"Where did it come from?"—is the most challenging. To answer these questions, we explored the literature presented here to gather relevant information on the function, development, and evolutionary relationships of this muscle and its homologs. The information that we gathered can be evaluated using a heuristic model known as a "decision tree" to guide students through their explorations of atypical morphology. The goal is to make explicit the nature of the information they need to pass each decision node and to understand the nature of any anatomic variant that they encounter. It is not meant to lead them to a predetermined "correct" answer, and the conclusions could vary depending on the nature and quality of the resources that students have available to complete their background research.

In this case, we formulated the "decision tree" as a graphic organizer, though there are other ways in which the process could be visualized and engaged in by students. Figure 5A shows the generalized decision tree that can be applied to any muscle with an atypical appearance.

Generalized Decision Tree for Assessing Atypical Muscles

Muscle appears to conform to a consistent general description of location, anatomic relationships, innervation, etc.

Yes

No

Muscle typically appears in other vertebrates

No

No homologous muscles located in other vertebrates

Muscle appears incomplete or unfinished

Yes

Muscle appears in common ancestor and all descendant branches (monophyly)

Muscle appears in several branches that do not derive it from common ancestor (paraphyly)

Muscle has consistent function

No

Yes

Muscle is fully functional in its own right

Yes

No

Function ancillary to that of other muscles

Atavism

Anomaly

Vestige

Remnant or Rudiment

(**A**)

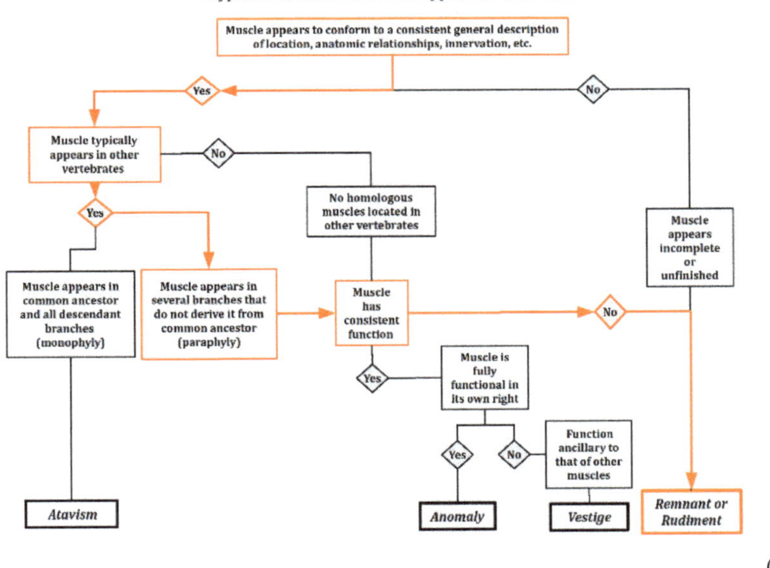

(**B**)

Figure 5. Decision Tree for assessing the likely sources of anatomic variations and areas for further inquiry. (**A**) Generalized form of decision tree showing nodes and links among pathways; (**B**) Application of decision tree to this particular example of M sternalis showing the realized decision pathway in orange.

For this example, aligning their anatomic findings for function, evolution, and development, students can use the criteria in the boxes to assess their inferences about the source of the M sternalis variants. As they pass each node in the decision tree, they will either (a) reach a decision as to the most likely type of the muscle variant; or (b) identify the areas in which more investigation is needed before making a final determination. For example, we discovered through the process the importance of the careful identification and preservation of neurovascular supply for unusual muscle variants. In future dissections—even in the absence of M sternalis—we learned the value of carefully examining the innervation of cutaneous and subcutaneous tissues before proceeding to deeper layers.

Figure 5B shows the application of the decision tree to the problem of the M sternalis in our lab. The pathway through the decision tree is highlighted in orange. With the data available to us, we concluded that this example of M sternalis most likely represents a remnant or a rudimentary form of some muscle, because there is no consistent function; even though the muscle appears to be well formed in our example, its morphology is not consistent across all the examples in the literature.

In our lab, the muscle in question was well formed with well-developed and firm attachments. However, it is clear to us that the presentation in our lab was only one of the variants known for M sternalis. Had we been presented with one of the other variants, we might have followed a different path through the decision tree, although most of the variants we see in the literature would still lead us to the same final conclusion.

The decision tree serves as a heuristic: a template for the process of inferring the nature of anatomic variants that present in the dissection lab. It allows students to view different aspects of nature of anatomic features and to focus any further research or discussion of alternative findings on specific issues related to any feature's function, evolutionary history, and development. The result is that students will have a more comprehensive appreciation for skeletal muscle structure and function that will enhance their understanding not only of atypical anatomic features, but of normal anatomy as well, and if they should palpate an unusual or unexpected muscle mass in their clinical practices, they will have a process for investigating and understanding the musculoskeletal variant that they have encountered.

Author Contributions: A.J.P. is Distinguished Lecturer Emeritus, Departments of Biological Sciences and Kinesiology at the University of Wisconsin-Milwaukee in Milwaukee, Wisconsin. He recently retired from over 30 years of teaching human anatomy and kinesiology. His research interests are in anatomic variation, functional morphology, and instructional strategies in life sciences. D.E.Z. is the founder and president of Little Red Pencil (https://littleredpencil.org/), an organization with the mission of increasing access to health care in Eastern Europe. He is currently a medical student at St George's University School of Medicine in Grenada, West Indies. E.K.J. is Associate Dean of Social Sciences at Southern New Hampshire University in Manchester, New Hampshire. Her research interests are within the social sciences, archaeology, and learning science. She was the laboratory instructor in the human gross anatomy course in the Doctoral Program in Physical Therapy at the UWM from 2012–2016. L.G. is a Physical Therapist with Advanced Physical Therapy and Sports Medicine in Marinette, Wisconsin. He specializes in post-operative orthopedic rehabilitation, joint replacements, and acute injury recovery, with a special interest in shoulder rehabilitation. J.T.M. received a Doctor of Physical Therapy degree from the University of Wisconsin–Milwaukee, where his interests in education and gross motor skill development led him to pursue clinical experience in school-based and sports-medicine physical therapy. He currently works as a physical therapist for the Milwaukee Public School District in Milwaukee, Wisconsin. N.W. is a physical therapist working for Froedtert and The Medical College of Wisconsin in Wauwatosa, Wisconsin. He graduated from the University of Wisconsin-Milwaukee's Doctor of Physical Therapy program in 2019. He practices outpatient physical therapy, treating a wide variety of impairments to improve his patients' quality of life. All authors have read and agreed to the published version of the manuscript.

Funding: This research received no external funding.

Acknowledgments: This work was reviewed by the Institutional Review Board for the University of Wisconsin-Milwaukee and ruled exempt under regulations applicable to human subjects and instructional improvement protocols. Some parts of the project were previously shared as poster presentations at a regional and a national conference of the Human Anatomy and Physiology Society. We thank James D Pampush of High Point University and Lyndsay M Barone of the Cold Spring Harbor DNA Learning Center for reviewing and commenting on earlier drafts of this paper. We are also grateful for the suggestions made by reviewers that helped us to improve the manuscript.

Conflicts of Interest: The authors declare no conflicts of interest.

References

1. Bergman, R.A.; Afifi, A.K.; Miyauchi, R. Illustrated Encyclopedia of Human Anatomic Variation. 2015. Available online: https://www.anatomyatlases.org/AnatomicVariants/AnatomyHP.shtml (accessed on 1 February 2020).
2. Bergman, R.A.; Thompson, S.A.; Afifi, A.K.; Saadeh, F.A. *Compendium of Human Anatomic Variation*; Urban & Schwarzenberg: Baltimore, MD, USA, 1988; ISBN 9780806725024.
3. Diogo, R.; Wood, B.A. *Comparative Anatomy and Phylogeny of Primate Muscles and Human Evolution*; CRC Press: Boca Raton, FL, USA, 2012; ISBN 9781578087679.
4. Platzer, W. *Color Atlas of Human Anatomy, Volume 1: Locomotor System*, 7th ed.; Thieme: New York, NY, USA, 2015; ISBN 9783135333076.
5. Diogo, R.; Abdala, V. *Muscles of Vertebrates: Comparative Anatomy, Evolution, Homologies, and Development*; CRC Press: Boca Raton, FL, USA, 2010; ISBN 9781578086825.
6. Turner, W. On the Musculus sternalis. *J. Anat. Physiol.* **1867**, *1*, 246–253. [CrossRef] [PubMed]
7. Stevenson, R.E.; Hall, J.G.; Goodman, R.M. *Human Malformations and Related Anomalies*; Oxford Monographs on Medical Genetics Nr 27; Oxford University Press: New York, NY, USA, 1993; Volume 2, ISBN 9780199386031.
8. Jelev, L.; Geirgiev, G.; Surchev, L. The sternalis muscle in the Bulgarian population: Classification of sternales. *J. Anat.* **2001**, *199*, 359–363. [CrossRef] [PubMed]
9. Ge, Z.F.; Tong, Y.L.; Zhu, S.Q.; Fang, X.; Zhuo, L.; Gong, X.Y. Prevalence and variance of the sternalis muscle: A study in the Chinese population using multi-detector CT. *Surg. Radiol. Anat. SRA* **2014**, *36*, 219–224. [CrossRef] [PubMed]
10. Eisler, P. *Die Muskeln des Stammes*; Gustav Fischer: Jena, Germany, 1912; ISBN 9780259142614.
11. Bailey, P.M.; Tzarnas, C.D. The sternalis muscle: A normal finding encountered during breast surgery. *Plast. Reconstr. Surg.* **1999**, *103*, 1189–1190. [CrossRef]
12. Demirpolat, G.; Oktay, A.; Bilgen, I.; Isayev, H. Mammographic features of the sternalis muscle. *Diagn. Interv. Radiol.* **2010**, *16*, 276–278.
13. Snosek, M.; Tubbs, R.S.; Loukas, M. Sternalis muscle, what every anatomist and clinician should know. *Clin. Anat.* **2014**, *27*, 866–884. [CrossRef]
14. Cabrolio, B. *Alphabēton Anatomikon: Hoc est Anatomes Elenchus Accuratissimus, Omnes Humani Corporis Partes, ea qua Solent Secari Methodo, Delineans*; Chouët: Geneva, Switzerland, 1603. Available online: https://www.e-rara.ch/gep_r/content/titleinfo/16295572 (accessed on 6 July 2020).
15. Cochon-DuPuy, J. Diverses observations anatomiques. In *Histoire de l'Académie Royale des Sciences*; Imprimerie Royale: Paris, France, 1726; pp. 26–27. Available online: https://gallica.bnf.fr/ark:/12148/bpt6k3588g/f34.image (accessed on 1 February 2020).
16. Albinus, B.S. *Historia Musculorum Hominis*; University of Bamberg: Bamberg, Germany, 1784; p. 262. ISBN 9781294089278. Available online: https://books.google.com/books?id=PtdTbQH9hlsC (accessed on 1 February 2020).
17. M'Whinnie, A.M. On the varieties in the muscular system of the human body. *Lond. Med. Gaz.* **1846**, *37*, 185–196.
18. Sharpey, W.; Thomson, A.; Schäfer, E.A. (Eds.) *Quain's Anatomy*, 8th ed.; Longmans, Green, and Co.: London, UK, 1876. Available online: https://babel.hathitrust.org/cgi/pt?id=njp.32101067214278 (accessed on 1 February 2020).
19. Gruber, W. Die Supernumären Brustmuskeln des Menschen. In *Mémoires de l'Academie Impériale des Sciences de St. Pétersbourg*, 8th ed.; L'Académie: St. Petersbourg, Russia, 1860. Available online: https://books.google.com/books?id=bZhFAAAAcAAJ&printsec=frontcover&dq=Gruber+Brustmuskeln+des+Menschen (accessed on 31 January 2020).
20. Sarikçioğlu, L.; Demirel, B.M.; Oğuz, N.; Uçar, Y. Three sternalis muscles associated with abnormal attachments of the pectoralis major muscle. *Anatomy* **2008**, *2*, 67–71. [CrossRef]
21. Natsis, K.; Totlis, T. A rare accessory muscle of the anterior thoracic wall. *Clin. Anat.* **2007**, *20*, 980–981. [CrossRef]
22. Hung, L.Y.; Lucaciu, O.C.; Wong, J.J. Back to the debate: Stemalis muscle. *Int. J. Morphol.* **2012**, *30*, 330–336. [CrossRef]

23. Arráez-Aybar, L.A.; Sobrado-Perez, J.; Merida-Velasco, J.R. Left musculus sternalis. *Clin. Anat.* **2003**, *16*, 350–354. [CrossRef] [PubMed]
24. Katara, P.; Chauhan, S.; Arora, R.; Saini, P.A. A unilateral rectus sternalis muscle: Rare but normal anatomical variant of anterior chest wall musculature. *J. Clin. Diagn. Res.* **2013**, *7*, 2665–2667. [CrossRef] [PubMed]
25. Kida, M.Y.; Izumi, A.; Tanaka, S. Sternalis muscle: Topic for debate. *Clin. Anat.* **2000**, *13*, 138–140. [CrossRef]
26. Patten, C.J. Right sternalis muscle with expanded fenestrated tendon. *J. Anat.* **1934**, *68*, 424–425.
27. Pillay, M.; Ramakrishnan, S.; Mayilswamy, M. Two cases of rectus sternalis muscle. *J. Clin. Diagn. Res.* **2016**, *10*, AD01–AD03. [CrossRef]
28. Vaithianathan, G.; Aruna, S.; Rajila, R.H.S.; Balaji, T. Sternalis "mystery" muscle and its clinical implications. *Ital. J. Anat. Embryol.* **2011**, *116*, 139–143.
29. Raikos, A.; Paraskevas, G.K.; Tzika, M.; Faustmann, P.; Triaridis, S.; Kordali, P.; Kitsoulis, P.; Brand-Saberi, B. Sternalis muscle: An underestimated anterior chest wall anatomical variant. *J. Cardiothorac. Surg.* **2011**, *6*, 73. [CrossRef]
30. O'Neil, M.N.; Folan-Curran, J. Case Report: Bilateral sternalis muscles with a bilateral pectoral anomaly. *J. Anat.* **1998**, *193*, 289–292. [CrossRef]
31. Williston, S.W. The sternalis muscle. *Proc. Acad. Nat. Sci. USA* **1889**, *41*, 38–41.
32. Carroll, M.A.; Lebron, E.M.; Jensen, T.E.; Cooperman, T.J. Chondroepitrochlearis and a supernumerary head of the biceps brachii. *Anat. Sci. Internat.* **2019**, *94*, 330–334. [CrossRef]
33. Tröbs, R.-B.; Gharavi, B.; Neid, M.; Cernaianu, G. Chondroepitrochlearis muscle—A Phylogenetic remnant with clinical importance. *Klin. Pädiatr.* **2015**, *227*, 243–246. [CrossRef]
34. Petto, A.J. The de–riving force of cladogensis. *Rep. Nat. Cent. Sci. Educ.* **1999**, *19*, 13.
35. Zanni, G.; Opitz, J.M. Annals of morphology. Atavisms: Phylogenetic Lazarus? *Am. J. Med. Gen. Part A* **2013**, *161A*, 2822–2835. [CrossRef] [PubMed]
36. Kirk, T.S. Sternalis muscle (in the living). *J. Anat.* **1925**, *59*, 192. [PubMed]
37. Hildebrand, M. *Analysis of Vertebrate Structure*, 3rd ed.; John Wiley & Sons: New York, NY, USA, 1982; ISBN 9780471825685.
38. Omura, A.; Ejima, K.-I.; Honda, K.; Anzai, W.; Taguchi, Y.; Koyabu, D.; Endo, H. Locomotion pattern and trunk musculoskeletal architecture among Urodela. *Acta Zool.* **2015**, *96*, 225–235. [CrossRef] [PubMed]
39. Getty, R. *Sisson and Grossman's The Anatomy of the Domestic Animals*; WB Saunders: Philadelphia, PA, USA, 1975; ISBN 9780721641027.
40. Swindler, D.R.; Wood, C.D. *An Atlas of Primate Gross Anatomy: Baboon, Chimpanzee, and Man*; University of Washington Press: Seattle, WA, USA, 1973; ISBN 9780898743210.
41. Osman Hill, W.C. *Primates: Comparative Anatomy and Taxonomy VIII. Cynopithecinae: Papio, Mandrillus, Theropithecus*; Interscience Publishers, Inc.: New York, NY, USA, 1970.
42. Osman Hill, W.C. *Primates: Comparative Anatomy and Taxonomy VII. Cynopithecinae: Cercocebus, Macaca, Cynopithecus*; Interscience Publishers, Inc.: New York, NY, USA, 1974; ISBN 9780852240144.
43. Hartman, C.G.; Strauss, W.L., Jr. (Eds.) *The Anatomy of the Rhesus Monkey*; Williams & Wilkins: Baltimore, MD, USA, 1933; ISBN 9780028457703.
44. Osman Hill, W.C. *Primates: Comparative Anatomy and Taxonomy II. Haplorhini: Tarsioidea*; Interscience Publishers, Inc.: New York, NY, USA, 1955; ISBN 9780471396994.
45. Osman Hill, W.C. *Primates: Comparative Anatomy and Taxonomy III. Pithecoidea: Platyrrhini (Families Hapalidae and Callimiconidae)*; Interscience Publishers, Inc.: New York, NY, USA, 1957.
46. Osman Hill, W.C. *Primates: Comparative Anatomy and Taxonomy IV. Cebidae Part A*; Interscience Publishers, Inc.: New York, NY, USA, 1960; ISBN 9780852241363.
47. Langworthy, O.R. The *Panniculus carnosus* in cat and dog and its genetical relation to the pectoral musculature. *J. Mammal.* **1924**, *5*, 49–63. [CrossRef]
48. Naldaiz-Gastesi, N.; Bahri, O.A.; López de Munain, A.; McCullagh, K.J.A.; Izeta, A. The *panniculus carnosus* muscle: An evolutionary enigma at the intersection of distinct research fields. *J. Anat.* **2018**, *233*, 275–288. [CrossRef]
49. Begun, D.R. How to identify (as opposed to define) a homoplasy: Examples from fossil and living great apes. *J. Hum. Evol.* **2007**, *53*, 559–572. [CrossRef]
50. Wolpert, L. *Developmental Biology: A Very Short Introduction*; Oxford University Press: New York, NY, USA, 2011; p. 50. ISBN 9780199601196.

51. McGready, T.A.; Quinn, P.J.; Fitzpatrick, E.S.; Ryan, M.T. *Veterinary Embryology*, 1st ed.; John Wiley and Sons: New York, NY, USA, 2006; ISBN 978-1405111478.
52. Shearman, R.M.; Burke, A.C. The lateral somitic frontier in ontogeny and phylogeny. *J. Exp. Zool.* **2009**, *312B*, 603–612. [CrossRef]
53. Pownall, M.E.; Gustafsson, M.K.; Emerson, C.P., Jr. Myogenic regulatory factors and the specification of muscle progenitors in vertebrate embryos. *Annu. Rev. Cell Dev. Biol.* **2002**, *18*, 747–783. [CrossRef]
54. Mekonen, H.K.; Hikspoors, J.P.J.M.; Mommen, G.; Köhler, E.; Lameres, W.H. Development of the ventral body wall in the human embryo. *J. Anat.* **2015**, *227*, 673–685. [CrossRef] [PubMed]
55. Pineault, K.M.; Wellik, D.M. *Hox* genes and limb musculoskeletal development. *Curr. Osteoporos. Rep.* **2014**, *12*, 420–427. [CrossRef] [PubMed]
56. Bannur, B.M.; Mallashetty, N.; Endigeri, P. An accessory muscle of pectoral region: A case report. *J. Clin. Diagn. Res.* **2013**, *7*, 1994–1995. [CrossRef] [PubMed]
57. Hasson, P. "Soft" tissue patterning: Muscles and tendons of the limb take their form. *Dev. Dyn.* **2011**, *240*, 1100–1107. [CrossRef]
58. Brand-Saberi, B. Genetic and epigenetic control of skeletal muscle development. *Ann. Anat.* **2005**, *187*, 199–207. [CrossRef]

Article

The Branching and Innervation Pattern of the Radial Nerve in the Forearm: Clarifying the Literature and Understanding Variations and Their Clinical Implications

F. Kip Sawyer [1,2,*], Joshua J. Stefanik [3] and Rebecca S. Lufler [1]

1 Department of Medical Education, Tufts University School of Medicine, Boston, MA 02111, USA; rebecca.lufler@tufts.edu
2 Department of Anesthesiology, Stanford University School of Medicine, Stanford, CA 94305, USA
3 Department of Physical Therapy, Movement and Rehabilitation Science, Bouve College of Health Sciences, Northeastern University, Boston, MA 02115, USA; j.stefanik@neu.edu
* Correspondence: ksawyer1@stanford.edu

Received: 20 May 2020; Accepted: 29 May 2020; Published: 2 June 2020

Abstract: Background: This study attempted to clarify the innervation pattern of the muscles of the distal arm and posterior forearm through cadaveric dissection. Methods: Thirty-five cadavers were dissected to expose the radial nerve in the forearm. Each muscular branch of the nerve was identified and their length and distance along the nerve were recorded. These values were used to determine the typical branching and motor entry orders. Results: The typical branching order was brachialis, brachioradialis, extensor carpi radialis longus, extensor carpi radialis brevis, supinator, extensor digitorum, extensor carpi ulnaris, abductor pollicis longus, extensor digiti minimi, extensor pollicis brevis, extensor pollicis longus and extensor indicis. Notably, the radial nerve often innervated brachialis (60%), and its superficial branch often innervated extensor carpi radialis brevis (25.7%). Conclusions: The radial nerve exhibits significant variability in the posterior forearm. However, there is enough consistency to identify an archetypal pattern and order of innervation. These findings may also need to be considered when planning surgical approaches to the distal arm, elbow and proximal forearm to prevent an undue loss of motor function. The review of the literature yielded multiple studies employing inconsistent metrics and terminology to define order or innervation.

Keywords: radial nerve; variation; order of innervation; posterior interosseous nerve; superficial branch of radial nerve; forearm

1. Introduction

Knowledge of the order of motor nerve branching from main nerve trunks into skeletal muscles is of clinical importance when evaluating motor and sensory deficits, treating nerve entrapment and predicting the timing of recovery after nerve injury [1]. Radial nerve entrapments, such as radial tunnel and posterior interosseous nerve syndromes, are common and result in reduced quality of life and ability to carry out activities of daily living [2].

Numerous sites of radial nerve entrapment have been identified in the literature—five for the posterior interosseous nerve alone—each resulting in a unique clinical presentation specific to the nerve's motor branching pattern [3]. It is imperative that clinicians have thorough knowledge of the radial nerve's course and branching in order to provide accurate and timely diagnosis.

Although some parametric studies of the radial nerve focused on the "order of innervation" in the forearm have been carried out, they have typically defined this term poorly and employed dissimilar metrics to represent it, making it difficult to compare studies [4,5]. Specifically, some authors have identified the order of motor branching from the radial nerve when determining the "order of innervation," while others have identified the order with which each muscle exhibits a motor entry point (the point at which a motor nerve first enters the muscle). Only a single study, carried out by Linell in 1921, documented both the branching order and motor entry order for the entire course of the radial nerve in the forearm, with findings indicating that those two orders are not always in agreement [6]. Thus, the methodological discrepancies present in modern studies prevent accurate comparisons and render them individually unable to present a complete picture of the radial nerve's motor branching pattern.

Beyond the knowledge of an archetypal "pattern" of the radial nerve and its branches in the forearm, an understanding of common variations is of particular clinical utility vis à vis preventing the iatrogenic loss of function or initial misdiagnosis of a lesion in a patient whose presenting symptomatology or underlying anatomy may be aberrant. Some existing studies have identified cases of innervation of the brachialis muscle by the radial nerve [4,7,8], as well as motor innervation by the superficial branch of radial nerve, which is often considered purely sensory [4,5,8]. These findings have thus far not been incorporated into most anatomical texts and remain relatively unknown.

This mixed methods study aimed to provide qualitative and quantitative data of both educational and clinical significance on the branching pattern and motor distribution of the radial nerve. The primary objectives of this study were to: (1) determine the branching order of the radial nerve in the distal arm and forearm; (2) determine the order of innervation of the radial nerve in the distal arm and forearm based on the distance to motor entry points; and (3) to identify the muscular territories of each branch of the radial nerve and quantify any variation thereof. The comprehensive measurement of both branching points and motor entry points set forth here provides a more complete and specific set of data than is currently available, as these parameters had previously been used independently to describe the nerve's anatomy or "order of innervation" [4,5]. These data will not only clarify existing reports in the literature, but also set forth an understanding of where and how branches divide from the radial nerve and reach their target structures.

2. Materials and Methods

2.1. Cadaveric Study Sample

This study utilized a sample of 35 embalmed cadavers housed in the gross anatomy laboratory at Tufts University School of Medicine that were concurrently being dissected as part of the medical and dental gross anatomy courses. One upper extremity from each cadaver was included. The selection of the laterality of each dissection was based on the fact that students performed a deep dissection of only one upper extremity, leaving the other one grossly intact and thus available for research purposes. As such, there was an inclusion of both right and left upper extremities. Cadaveric research of this nature does not require the Institutional Review Board or ethics approval at Tufts University.

2.2. Measurement of Radial Nerve

The 35 upper extremities were dissected initially to reveal the superficial musculature and then further to distally expose the radial nerve from the level of the spiral groove of the humerus proximally to the arcade of Frohse (supinator arch). With the extremity in anatomical position, the medial and lateral humeral epicondyles were palpated and a line was drawn bisecting them (the transepicondylar line, TEL). A pin was placed through the radial nerve at the point where it crossed the TEL to both fix it in place and serve as a consistent marker for measurement. Each muscular branch of the nerve was identified and the distance from the TEL to the nerve's branch point from the radial nerve ("branch distance") was measured to the nearest one-hundredth of a millimeter (0.01 mm) using a digital caliper (Neiko Tools®, Taipei, Taiwan). The branch points proximal to the TEL were recorded as negative values and those distal were recorded as positive. The length of each branch (a straight line from its exit from the radial nerve to its motor entry point—MEP—at the muscle) was then measured in the same fashion. The branch distance and branch length were then summed to determine the "MEP distance." This was first performed for brachialis (B; if innervated supinator (S). The supinator was then reflected to allow for the dissection of the nerves supplying extensor carpi ulnaris (ECU), extensor digitorum and digiti minimi (ED, EDM), abductor pollicis longus (APL), extensor pollicis longus and brevis (EPL, EPB) and extensor indicis (EI).

Previous studies have used differing and often poorly defined terminology when discussing the distal branches of the radial nerve. In this study the deep branch of the radial nerve (DBRN) was defined as the nerve that remains after the superficial branch of the radial nerve (SBRN) branches from the radial nerve, while the posterior interosseous nerve (PIN) was defined as the continuation of DBRN distal to the distal border of the supinator. This is consistent with the definition used by Moore et al. in *Clinically Oriented Anatomy* [9].

2.3. Analysis of Radial Nerve Branches

The most common source of innervation for each muscle was determined based on the frequency with which each named branch of the radial nerve supplied it. A "most common branching order" was determined based on the mean branch distance for each muscle across all specimens. A "most common MEP order" was then determined based on the mean distance to the first MEP for each muscle. For both of these variables a one-way analysis of variance (ANOVA) followed by a post-hoc Tukey's test was used to evaluate the differences between the muscles. Significance was set to $p < 0.05$. Studies since Linell's in 1921 have used either one or the other of these two variables when defining the "order of innervation"; this study notably investigates both as we believe they each contribute to the conceptualization of an overall order of innervation and serve unique purposes in different clinical contexts.

3. Results

3.1. Branching Order of the Radial Nerve in the Forearm

Thirty-five cadavers, 14 of which were female and 21 of which were male, with a mean age of 79.9 ± 11.5, were included in this study. The muscular nerve branching order was determined based on the mean distance from the TEL to the first branch exiting off the main trunk to serve each muscle (branch distance, Table 1). The branching order is presented as follows, with (*) representing a significant difference in the mean branch distance ($p \leq 0.05$): brachialis, * brachioradialis, * ECRL,* ECRB, supinator,* ED, ECU, APL, EDM,* EPB,* EPL, EI. The mean branch distance was not statistically significantly different between the ECRB and the supinator; the ED and ECU; any of the ECU, APL and EDM; or the EI and EPL. The ECRB's branch arose proximal to that of supinator in 27 specimens (77.1%); ED's was proximal to ECU's in 24 (68.6%); ECU's was proximal to APL's in 33 (94.3%); ECU's was proximal to EDM's in 28 (80.0%); and EDM's was proximal to APL's in 18 (51.4%). The branch to the EPL arose proximal to that to the EI in 10 (28.6%) and the two arose co-terminally in 22 (62.9%).

3.2. Motor Entry Point Order

The distance from the TEL to the most proximal entry point of a nerve into each muscle (MEP distance) is depicted in Table 2. That order is as follows, with (*) representing a significant difference in the mean MEP distance ($p \leq 0.05$): brachialis, * brachioradialis, * ECRL, * supinator, ECRB,* ED, ECU, EDM, APL,* EPL, EPB,* EI. The mean distance to the MEP was not statistically significant between the supinator and ECRB; ED and ECU; EDM and APL; or the EPL and EPB. The supinator's MEP was proximal to the ECRB's in 27 (77.1%); ED's was proximal to ECU's in 25 (71.4%); EDM's was proximal to APL's in 24 (68.6%); EPL's was proximal to EPB's in 20 (57.1%); EPB's was proximal to EI's in 23 (65.7%).

Table 1. Branching order, based on the mean distance in mm along the trunk from the transepicondylar line to the first branch to a muscle. p values are listed for the comparison of a mean to the one below it. p values comparing any muscle to another muscle not adjacent to it on this table were all <0.05, except for the extensor carpi ulnaris (ECU) and extensor digiti minimi (EDM) ($p = 0.1281$).

Muscle	N	Mean	Standard Deviation	p (To Next Muscle)
Brachialis	21	−66.41	19.63	<0.0001 *
Brachioradialis	35	−40.37	12.08	0.0036 *
Extensor carpi radialis longus	35	−26.02	10.53	<0.0001 *
Extensor carpi radialis brevis	35	17.88	18.29	0.9553
Supinator	35	22.78	11.56	<0.0001 *
Extensor digitorum	35	81.90	12.03	0.9996
Extensor carpi ulnaris	35	84.60	10.60	0.1337
Abductor pollicis longus	35	94.92	15.26	1.0
Extensor digiti minimi	35	94.99	15.99	<0.0001 *
Extensor pollicis brevis	35	112.44	21.28	0.0271 *
Extensor pollicis longus	35	124.75	16.48	1.0
Extensor indicis	35	126.84	16.66	Not Applicable

* Indicates significant difference in mean branching distance between the listed muscle and the muscle below it (i.e., more distally) on the chart.

3.3. Motor Innervation Territories

Descriptive statistics of muscular territories are shown in Table 3. Brachialis received a branch from the radial nerve proper in 21 specimens (60%). Brachioradialis and ECRL were always innervated by the radial nerve proper, although brachioradialis also received innervation by a much more substantial branch from the SBRN in one specimen (2.9%) (Figure 1); the ECRL was solely supplied by radial nerve in all specimens (100%). The ECRB received sole innervation from the radial nerve in nine specimens (25.7%), DBRN in 16 (45.7%) and the SBRN in eight (22.9%, Figure 2); it received dual innervation from the radial nerve and DBRN in one specimen (2.9%). Supinator was generally innervated by DBRN (89%), but occasionally was served more proximally by the radial nerve (11%). The ED was innervated by PIN in 30 specimens (85.7%), DBRN in four (11.4%) and by the two nerves together in one (2.9%) The following muscles were generally innervated by PIN: ECU (82.9%), EDM (97.1%), APL (91.4), EPB (97.1%); they were otherwise supplied by DBRN (17.1%, 2.9%, 8.6%, 2.9%, respectively). EPL and EI were both exclusively innervated by PIN in all 35 specimens (100%).

3.4. Landmarks

With the TEL located at 0.00 mm in all the specimens, the division of the radial nerve into the superficial branch of the radial nerve (SBRN) and deep branch of the radial nerve (DBRN) occurred at a mean of 9.79 ± 12.2 mm distally. The arcade of Frohse (supinator arch) occurred at 42.3 ± 7.94 mm. The distal margin of the supinator, where the DBRN becomes the PIN, occurred at 78.6 ± 8.67 mm from the TEL.

Table 2. Motor entry point order, based on the mean distance in mm from the transepicondylar line to the entry into the muscle (sum of branch distance, above and branch length). *p* values are listed for the comparison of a mean to the one below it. *p* values comparing any muscle to another muscle not adjacent to it on this table were all <0.05.

Muscle	N	Mean	Standard Deviation	*p* (To Next Muscle)
Brachialis	21	−52.22	23.19	<0.0001 *
Brachioradialis	35	−20.22	17.11	<0.0001 *
Extensor carpi radialis longus	35	5.10	10.04	<0.0001 *
Supinator	35	47.67	8.42	0.4593
Extensor carpi radialis brevis	35	57.13	19.60	<0.0001 *
Extensor digitorum	35	94.24	13.68	0.9896
Extensor carpi ulnaris	35	99.08	13.28	0.0017 *
Extensor digiti minimi	35	116.33	17.39	0.8526
Abductor pollicis longus	35	123.38	13.15	<0.0001 *
Extensor pollicis longus	35	147.56	21.91	0.5003
Extensor pollicis brevis	35	154.22	22.24	0.0062 *
Extensor indicis	35	163.47	20.40	Not Applicable

* Indicates significant difference in mean motor entry point distance between the listed muscle and the muscle below it (i.e., more distally) on the chart.

Table 3. Percentage distribution of each muscle among the branches of the radial nerve (e.g., muscular territories). Note: percentages for the brachioradialis (BR), the extensor carpi radialis brevis (ECRB) and the extensor digitorum (ED) sum to greater than 100% as certain specimens had dual innervation to these muscles.

Muscle	Radial Nerve	Superficial Branch of the Radial Nerve	Deep Branch of the Radial Nerve	Posterior Interosseous Nerve
Brachialis	60			
Brachioradialis	100	2.86		
Extensor carpi radialis longus	100			
Extensor carpi radialis brevis	28.57	25.71	48.57	
Supinator	11.43		88.57	
Extensor digitorum			14.29	88.57
Extensor carpi ulnaris			17.14	82.86
Extensor digiti minimi			2.86	97.14
Abductor pollicis longus			8.57	91.43
Extensor pollicis brevis			2.86	97.14
Extensor pollicis longus				100
Extensor indicis				100

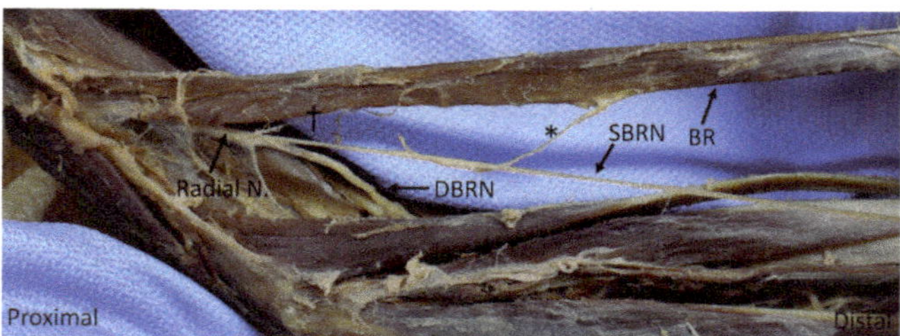

Figure 1. Nerve to the brachioradialis (BR) arising from the superficial branch of a radial nerve (*), with an additional smaller branch arising from the radial nerve proper (†). BR: brachioradialis; SBRN: superficial branch of the radial nerve; DBRN: deep branch of the radial nerve.

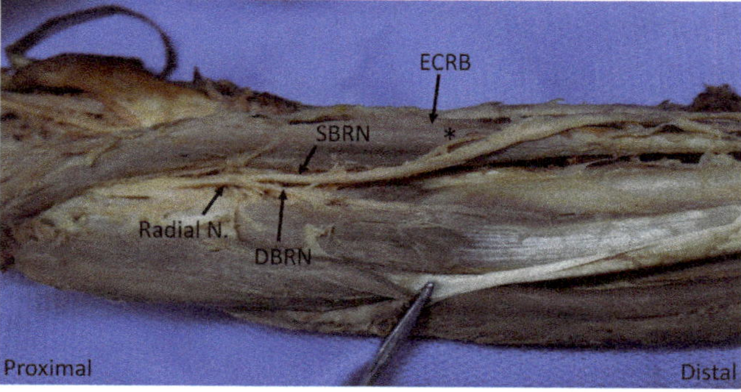

Figure 2. Nerve to extensor carpi radialis brevis (*) arising from the superficial branch of the radial nerve. ECRB: extensor carpi radialis brevis; SBRN: superficial branch of radial nerve; DBRN: deep branch of radial nerve.

4. Discussion

4.1. Branching and Motor Entry Point Orders

The order with which branches diverged from the main trunk of the radial nerve to serve the forearm extensors exhibited an overarching and relatively predictable pattern in most specimens. Deviation from the norm was notable in specific, consistent pairs (and one triplet) of muscles—often of the deep layer—whose branches typically exited near one another and thus were prone to swapping. This is evidenced by the statistically insignificant difference in the mean branch point distances between the ECRB and supinator; ED and ECU; ECU and APL and EDM; and the EI and EPL. There was similar variation in the MEP order, with statistically insignificant differences between the ECRB and supinator; ED and ECU; EDM and APL; and the EPL and EPB.

The overall MEP order found here is consistent with that described by Abrams et al., who used this metric to define the "order of innervation" [4]. Branovacki et al. produced a descriptive (i.e., not quantitative) account of the most common branch order in sixty specimens that agrees with the overall findings on the branching order here [5]. Lastly, although Mazurek and Shin did not explicitly report how they defined their "most commonly accepted order of innervation," their list does agree with this investigation's findings on the branching order [2].

Knowledge of the order in which muscles are innervated by the radial nerve, along with the length of individual branches, has important clinical implications. The shortest distance along a nerve trunk and the subsequent muscular branch (represented in this study by the MEP distance) is an important predictor of the order and timing of re-innervation (and thus the recovery of individual muscle function) after nerve injury [1,4]. Furthermore, the typical order in which nerves branch to serve muscles (here represented by branch distance) is an important factor to consider when localizing lesions in cases of nerve entrapment [2]. This information is particularly helpful when used in conjunction with knowledge of the locations of structures which are known to cause entrapment, such as the arcade of Frohse, as the relative location of these structures to individual motor branches can inform specific localization. Lastly, when understood together, branch order and MEP distance can aid in conceptualizing the overall course of the motor branches, which is important when dissecting in the vicinity of the nerve during surgery or planning for nerve transfers [8,10].

4.2. Innervation of Brachialis by Radial Nerve

This study noted the partial innervation of brachialis by the radial nerve in 60% of specimens, with the branch arising from the radial nerve at a mean of 66.4 ± 19.6 mm proximal to the TEL, in the plane between the brachialis and brachioradialis muscles. This is consistent with the findings of previous studies in Western populations that have documented rates of 50% [4], 65% [8] and 67% [7], and lower than the rates of 81.6% [11] and 100% [12] documented in Asian populations; Blackburn et al. hypothesized that part of this discrepancy may be due interracial differences [7]. As this branch from the radial nerve—present in potentially two-thirds of patients—travels in the plane between the brachialis and brachioradialis, care should be taken when surgically dissecting these two muscles apart (such as in an anterior approach to the humerus) to prevent the denervation of the lateral portion of the brachialis [7,11].

A survey of the six most commonly used preclinical anatomical texts at Tufts University School of Medicine—*Atlas of Human Anatomy*, 6th edition [13]; *Clinically Oriented Anatomy*, 7th edition [9]; *Essential Clinical Anatomy*, 5th edition [14]; *Grant's Atlas of Anatomy*, 13th edition [15]; *Gray's Anatomy for Students*, 3rd edition [16]; *Atlas of Anatomy*, 2nd edition [17]—yielded no mention of this common source of innervation. Given that the results of this study support the results of previous studies that the brachialis muscle is partially innervated by the radial nerve at a greater than 50% incidence rate, it can be argued that this should be regularly included in all textbooks and atlases, as well as in medical student anatomy curricula.

4.3. Muscular Innervation by Superficial Branch of Radial Nerve

The superficial branch of the radial nerve was found to provide sole innervation to ECRB in 25.7% of specimens in this study (Figure 2). The breakdown of the innervation source for the ECRB was remarkably consistent with that found by Abrams et al., who documented innervation rates of 25% by SBRN, 30% by radial nerve and 45% by what the authors termed PIN (which we would refer to as the DBRN as it has not yet reached the distal border of supinator; see below) [4]. These findings are noteworthy given that there exists a commonly held notion that the SBRN is a purely sensory nerve and serves no motor function. As above, none of the six textbooks surveyed discussed this common variant, and many reinforced the teaching that SBRN is sensory-only.

Furthermore, the current study sample also included one specimen with what appeared to be the primary innervation of the brachioradialis by SBRN (with an additional, much less substantial branch from the radial nerve, Figure 1 above). A search of the existing literature revealed no other studies that have reported such an anomaly. A limitation to this study is that it cannot confirm that this nerve from SBRN carried motor fibers, although its bulky appearance and the lack of other significant nerve supply was quite suggestive of this.

Knowledge of the potential for innervation of either the ECRB or brachioradialis muscles by the SBRN would be of significant importance in many surgical settings. Given the SBRN's very superficial location under the skin and lack of robust protection, it is susceptible to iatrogenic injury during various surgical approaches, including the commonly performed radial forearm free flap for the reconstruction of tissue defects, where overlying tissue is harvested while the SBRN is ideally left intact [18,19]. It is well known that injury to the distal portion of the nerve can result in intractable pain or neuroma formation [20], but there has been less focus on the potential for proximal injury, let alone the possible consequence of muscular denervation that such injury could entail. Although the SBRN is more proximally protected thanks to the overlying mobile wad, it is still susceptible to injury or entrapment due to surgical dissection, post-surgical edema or scarring, or—in a case of parathyroid autotransplantation followed post-operatively by one of the authors (F.K.S.)—delayed transplanted gland hypertrophy. The knowledge that SBRN has a motor component in nearly a third of the studied specimens should prompt heightened concern for the preservation of the nerve if at all possible.

4.4. Inconsistency of Nomenclature and Territories

Through the course of reviewing existing literature, it became apparent that a disparity exists in the naming conventions for the DBRN and its successor, PIN, between clinicians and anatomists. Most notably, literature authored by physicians typically did not distinguish DBRN from PIN at all. Instead, the nerve that remains after the exit of SBRN from the radial nerve was referred to exclusively as the PIN by multiple authors [2,4,5]. This is in contrast to anatomical textbooks; for instance, *Clinically Oriented Anatomy* references the division of the radial nerve into the SBRN and DBRN, and describes the course of the latter as follows:

"the deep branch of the radial nerve, after it pierces the supinator, runs in the fascial plane between superficial and deep extensor muscles in close proximity to the posterior interosseous artery; it is usually referred to as the posterior interosseous nerve" [9].

Terminologia Anatomica agrees, recognizing a "deep branch" of the radial nerve that terminates in the "posterior interosseous nerve" [21]. It is unclear as to why clinicians deviate from this paradigm and combine DBRN into PIN; regardless, it leads to lack of clarity and represents a discrepancy between basic science and clinical terminology that would ideally be reconciled.

5. Conclusions

Although the radial nerve does exhibit a noteworthy degree of variability in the posterior forearm, in terms of both branch territories and order of innervation, this investigation identified a pattern that was overall quite predictable. The results here serve to provide a more cohesive view of the morphologic parameters and patterning of the radial nerve and its motor branches in the distal arm and forearm than has been offered in the existing literature and to bring to further light some notable variations have often been overlooked. Given that injury to, or disease of, the radial nerve can present significant functional limitations that negatively affect quality of life [22], it is imperative that the relevant clinicians understand the nerve's anatomy and its potential for variation in thorough detail. These findings will ideally help to inform clinicians in their approach to the examination and diagnosis of radial nerve pathology, as well as in the safe performance of surgical dissection in the nerve's vicinity.

Author Contributions: Conceptualization, F.K.S., R.S.L.; methodology, F.K.S., R.S.L.; software, J.J.S.; formal analysis, F.K.S., R.S.L., J.J.S.; data curation, J.J.S.; writing—original draft preparation, F.K.S.; writing—review and editing, R.S.L., J.J.S. All authors have read and agreed to the published version of the manuscript.

Funding: This research received no external funding.

Acknowledgments: The authors would like to acknowledge Anne Agur for her assistance in revising this manuscript.

Conflicts of Interest: The authors declare no conflict of interest.

References

1. Sunderland, S. Metrical and non-metrical features of the muscular branches of the radial nerve. *J. Comp. Neurol.* **1946**, *85*, 93–111. [CrossRef] [PubMed]
2. Mazurek, M.T.; Shin, A.Y. Upper extremity peripheral nerve anatomy: Current concepts and applications. *Clin. Orthop. Relat. Res.* **2001**, *383*, 7–20. [CrossRef] [PubMed]
3. Dang, A.C.; Rodner, C.M. Unusual Compression Neuropathies of the Forearm, Part I: Radial Nerve. *J. Hand Surg.* **2009**, *34*, 1906–1914. [CrossRef] [PubMed]

4. Abrams, R.A.; Ziets, R.J.; Lieber, R.L.; Botte, M.J. Anatomy of the radial nerve motor branches in the forearm. *J. Hand Surg.* **1997**, *22*, 232–237. [CrossRef]
5. Branovacki, G.; Hanson, M.; Cash, R.; Gonzalez, M. The innervation pattern of the radial nerve at the elbow and in the forearm. *J. Hand Surg.* **1998**, *23*, 167–169. [CrossRef]
6. Linell, E.A. The Distribution of Nerves in the Upper Limb, with reference to Variabilities and their Clinical Significance. *J. Anat.* **1921**, *55*, 79–112. [PubMed]
7. Blackburn, S.C.; Wood, C.P.J.; Evans, D.J.R.; Watt, D.J. Radial nerve contribution to brachialis in the UK Caucasian population: Position is predictable based on surface landmarks. *Clin. Anat.* **2007**, *20*, 64–67. [CrossRef] [PubMed]
8. Cho, H.; Lee, H.-Y.; Gil, Y.-C.; Choi, Y.-R.; Yang, H.-J. Topographical anatomy of the radial nerve and its muscular branches related to surface landmarks. *Clin. Anat.* **2013**, *26*, 862–869. [CrossRef] [PubMed]
9. Moore, K.L.; Dalley, A.F.; Agur, A.M.R. *Clinically Oriented Anatomy*, 7th ed.; Wolters Kluwer Health/Lippincott Williams & Wilkins: Philadelphia, PA, USA, 2013; p. 764.
10. Abrams, R.A.; Brown, R.A.; Botte, M.J. The superficial branch of the radial nerve: An anatomic study with surgical implications. *J. Hand Surg.* **1992**, *17*, 1037–1041. [CrossRef]
11. Mahakkanukrauh, P.; Somsarp, V. Dual innervation of the brachialis muscle. *Clin. Anat.* **2002**, *15*, 206–209. [CrossRef]
12. Ip, M.C.; Chang, K.S.F. A study on the radial supply of the human brachialis muscle. *Anat. Rec.* **1968**, *162*, 363–371. [CrossRef]
13. Netter, F.H. *Atlas of Human Anatomy*, 6th ed.; Elsevier Health Sciences: Amsterdam, The Netherlands, 2014.
14. Moore, K.L.; Agur, A.M.R.; Dalley, A.F. *Essential Clinical Anatomy*, 5th ed.; Wolters Kluwer Health: Waltham, MA, USA, 2014.
15. Agur, A.M.R.; Dalley, A.F. *Grant's Atlas of Anatomy*, 13th ed.; Wolters Kluwer Health/Lippincott Williams & Wilkins: Philadelphia, PA, USA, 2009.
16. Drake, R.; Vogl, A.W.; Mitchell, A.W.M. *Gray's Anatomy for Students*, 3rd ed.; Elsevier Health Sciences: Amsterdam, The Netherlands, 2014.
17. Gilroy, A.M.; MacPherson, B.R.; Ross, L.M. *Atlas of Anatomy*, 2nd ed.; Thieme: New York, NY, USA, 2012.
18. Folberg, C.R.; Ulson, H.; Scheidt, R.B. The superficial branch of the radial nerve: A morphologic study. *Rev. Bras. De Ortop.* **2009**, *44*, 69–74. [CrossRef]
19. Haugen, T.W.; Cannady, S.B.; Chalian, A.A.; Shanti, R.M. Anatomical Variations of the Superficial Radial Nerve Encountered during Radial Forearm Free Flap Elevation. *ORL J. Otorhinolaryngol. Relat. Spec.* **2019**, *81*, 155–158. [CrossRef] [PubMed]
20. Dellon, A.L.; Mackinnon, S.E. Susceptibility of the Superficial Sensory Branch of the Radial Nerve to Form Painful Neuromas. *J. Hand Surg.* **1984**, *9*, 42–45. [CrossRef]
21. Federative Committee on Anatomical Terminology. *Terminologia Anatomica: International Anatomical Terminology*; Thieme: Stuttgart, Germany, 1998.
22. Stonner, M.M.; Mackinnon, S.E.; Kaskutas, V. Predictors of Disability and Quality of Life With an Upper-Extremity Peripheral Nerve Disorder. *Am. J. Occup.* **2017**, *71*, 7101190050p1–7101190050p8. [CrossRef] [PubMed]

Article

Anatomic and Histological Features of the Extensor Digitorum Longus Tendon Insertion in the Proximal Nail Matrix of the Second Toe

Patricia Palomo-López [1], Marta Elena Losa-Iglesias [2], Ricardo Becerro-de-Bengoa-Vallejo [3], David Rodríguez-Sanz [3], Cesar Calvo-Lobo [3,*], Jorge Murillo-González [4] and Daniel López-López [5]

1 University Center of Plasencia, Faculty of Podiatry, Universidad de Extremadura, 10600 Badajoz, Spain; patibiom@unex.es

2 Faculty of Health Sciences, Universidad Rey Juan Carlos, 28922 Alcorcon, Spain; marta.losa@urjc.es

3 School of Nursing, Physiotherapy and Podiatry, Universidad Complutense de Madrid, 28040 Madrid Spain; ribebeva@ucm.es (R.B.-d.B.V.); davidrodriguezsanz@ucm.es (D.R.-S.)

4 Department of Human, Anatomy and Embryology, Faculty of Medicine, Madrid Complutense University, 28040 Madrid, Spain; jmurillo@med.ucm.es

5 Research, Health and Podiatry Group. Department of Health Sciences, Faculty of Nursing and Podiatry, Universidade da Coruña, 15403 Ferrol Spain; daniellopez@udc.es

* Correspondence: cescalvo@ucm.es; Tel.: +34-913-941-532

Received: 18 February 2020; Accepted: 5 March 2020; Published: 7 March 2020

Abstract: Background: Anatomic and histological landmarks of the extensor digitorum longus (EDL) tendon insertion in the proximal nail matrix may be key aspects during surgery exposure in order to avoid permanent nail deformities. Objective: The main purpose was to determine the anatomic and histological features of the EDL's insertion to the proximal nail matrix of the second toe. Methods: A sample of fifty second toes from fresh-frozen human cadavers was included in this study. Using X25-magnification, the proximal nail matrix limits and distal EDL tendon bony insertions were anatomically and histologically detailed. Results: The second toes' EDLs were deeply located with respect to the nail matrix and extended superficially and dorsally to the distal phalanx in all human cadavers. The second toe distal nail matrix was not attached to the dorsal part of the distal phalanx base periosteum. Conclusions: The EDL is located plantar and directly underneath to the proximal nail matrix as well as dorsally to the bone. The proximal edge of the nail matrix and bed in human cadaver second toes are placed dorsally and overlap the distal EDL insertion. These anatomic and histological features should be used as reference landmarks during digital surgery and invasive procedures.

Keywords: anatomy and histology; foot; nails; nail matrix; toe joint; tendons; toe phalanges; nail deformity; anatomic landmarks

---

1. Introduction

Toe-tip disorders seem to be common and often affect the nail matrix and plate of the second toe [1]. Nail matrix and plate surgeries of the second toe distal phalanx may be performed secondary to trauma, infection, neoplasm, iatrogenic injuries, degenerative pathologies, and self-inflicted injuries [1,2].

Thus, surgeries for nail bed reconstruction may require a deep understanding of the anatomy and histology of the nail matrix as well as the relationship between the nail and surrounding tissues. The matrix may frequently develop injuries and deformities secondary to surgeries of the second toe distal phalanx and its nail plate. In addition, the nail matrix may be considered as the main origin of the nail plate. A higher potential risk is presented for scarring procedures in the proximal nail bed compared with the distal nail bed [3]. Finally, nail deformities may be produced secondary to the aforementioned conditions [4].

Anatomic and histological landmarks of the matrix proximal edge may be key aspects during surgery exposure in order to avoid permanent nail deformities. Several studies have evaluated the nail anatomy [4–11] as well as the flexor and extensor hallucis brevis and abductor and adductor hallucis tendon insertions of the hallux proximal phalanx base with surgical implications [12]. Indeed, the anatomic and histologic relationship between the proximal nail matrix and the distal tendons insertion has already been detailed for the distal extensor pollicis longus tendon of the fingernail matrix in the thumb [13], as well as the extensor hallucis longus tendon insertion into the proximal nail matrix of the great toe [14]. Nevertheless, to our knowledge, this anatomic and histologic relationship between the proximal nail matrix and the extensor digitorum longus (EDL) tendon insertion has not yet been described in the lesser toes. Thus, the purpose of this study was to determine the anatomic and histological features of the EDL insertion to the proximal nail matrix of the second toe.

2. Materials and Methods

2.1. Sample

A total sample of 50 2nd toes from fresh-frozen human cadavers, with an age older than 18 years old and without any history of digital traumas, was included for the present study. An age mean of 62.5 years with a range from 45 to 80 years old was obtained from the total sample, including 21 women and 29 men. These 2nd toes were obtained from 26 left feet and 24 right feet from 50 cadavers. Some toes were excluded due to the presence of digital deformities. The human cadavers proceeded from the Human Anatomy and Embriology I Department of the School of Medicine from the Complutense University of Madrid (Spain). This research was carried out according to The Strengthening the Reporting of Observational Studies in Epidemiology criteria [15].

2.2. Ethical Considerations

All experimental protocols were approved (1 March 2018) by the Ethical Committee of the Rey Juan Carlos University (Spain), with approval code 0801201800818. Furthermore, all methods were carried out according to the relevant regulations and guidelines. Consent for the present study was obtained from the Human Anatomy and Embriology I Department of the School of Medicine from the Complutense University of Madrid (Spain), which included informed consent as part of the cadaver donation process.

2.3. Procedure

A longitudinal incision of the dorsal skin was carried out along the shaft of the distal phalanx extending through the eponychium. In addition, two radial incisions were performed proximal to the junctions of the nail fold and nail walls [16,17]. Furthermore, skin flaps were retracted to expose the matrix dorsal part and the complete nail plate length.

Toenails were removed atraumatically, preserving the nail bed and matrix as well as the eponychium. These procedures were achieved by firstly loosening the nail plate from the nail matrix using a fine elevator. Nail plates were elevated and sharply excised from side to side. Nail plate removal enabled the accurate inspection of the nail matrix borders (Figure 1a).

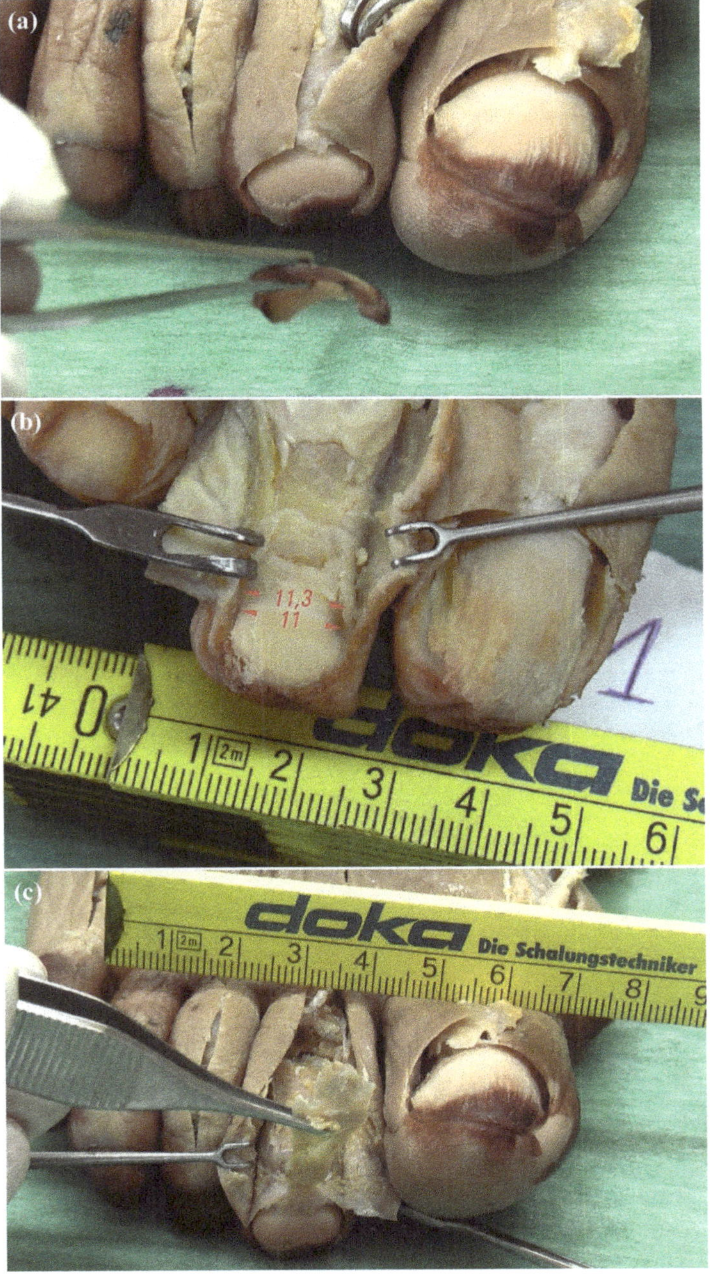

Figure 1. Anatomical study procedure: (**a**) The nail plate was gently elevated and sharply excised from side to side; (**b**) Color and texture of the nail matrix and bed (Red arrows and numbers showed the nail matrix width in mm at the proximal and distal nail parts); (**c**) Exposure of the distal insertion of the extensor digitorum longus (EDL) tendon at the base of distal phalanx until the proximal nail matrix fold was reached.

Proximal nail matrix edges were determined by visual inspection of the proximal origin location of the nail plate. Proximal nail plate origins were delimited as the proximal matrix edges. Afterwards, the color and texture of the nail matrix and bed were observed. Indeed, the nail matrix presented a white color, while nail beds presented a red color (Figure 1b).

Proximal matrix edges were identified with respect to the distal limits of the EDL tendon at the distal bony insertion by both X25-magnification optical microscope and visual inspection (Figure 1c).

Distal EDL tendon insertions were landmarked by visual inspection and confirmed by histological cross-sections analyses in the pathology laboratory. Standard longitudinal sections were stained by Tetrachrome VOF-III GS stains (light green SF/or fast green FCF-methyl blue-Orange G-acid fuchsin).

2.4. Outcome Measurements

Nail matrix proximal edges were stated in all cadavers. Distal EDL insertions were clearly identified in all cadavers. Thus, anatomical landmarks were macroscopically measured and delimited, as described below, according to the following variables. Variable-1: Matrix width measurements were carried out from the lateral to the medial side at the most proximal part of the 2nd toe. Variable-2: The same measurements were carried out at the distal aspect of the matrix, where colors changed with the nail bed. Variable-3: Matrix lengths were determined from proximal to distal sides along the shaft of the 2nd toe. Variable-4: Nail bed lengths were measured as above from where the color changed to the bed end. Variable-5: EDL tendon widths were determined from the lateral to the medial side at the distal interphalangeal joint line.

2.5. Statistical Analysis

Outcome measurements were detailed as the mean, standard deviation (SD), upper and lower limits of the 95% confidence interval (95% CI) and minimum and maximum values, as well as the median. All analyses were carried out with SPSS 19.0 software (Chicago, IL, USA). Data normality was assessed by the Kolmogorov–Smirnov test. In order to compare outcome measurements for sex and lower limb side distributions, the Student *t* test for independent samples and the Mann–Whitney U test were used for parametric and non-parametric data, respectively.

3. Results

The dimensions of the matrix, bed and EDL tendon of the second toes ($n = 50$) are presented in Table 1. Nail matrix widths were larger at the distal nail parts (8.80 ± 1.86 mm) than the proximal nail parts (8.07 ± 1.53 mm). Nail bed lengths (5.68 ± 1.40 mm) were larger than nail matrix lengths (2.37 ± 0.29 mm). In addition, the EDL tendon widths showed a mean ± SD of 2.49 ± 0.66 mm at the distal interphalangeal joint line.

Table 1. Dimensions of the matrix, bed and EDL tendon of the second toes ($n = 50$).

Outcome Measurements (mm; $n = 50$)	Mean ± SD	95% CI (LL-UL)	Median	Minimum	Maximum
1. Nail matrix width at the proximal nail part (mm)	8.07 ± 1.53	(8.54–9.39)	8.00	5.50	11.30
2. Nail matrix width at the distal nail part (mm)	8.80 ± 1.86	(8.28–9.31)	8.40	5.80	12.00
3. Length of matrix (mm)	2.37 ± 0.29	(2.28–2.45)	2.40	1.60	2.80
4. Length of bed (mm)	5.68 ± 1.40	(5.26–6.06)	5.40	3.30	8.70
5 Tendon width (mm)	2.49 ± 0.66	(2.30–2.68)	2.60	1.50	4.0

Abbreviations: CI, confidence interval; EDL, extensor digitorum longus; mm, millimeters; LL, lower limit; UL, upper limit.

According to sex comparisons presented in Table 2, statistically significant differences ($p < 0.001$) were shown for the nail matrix width at the proximal nail part showing a larger width for men compared to women. The rest of the outcome measurements did not show any statistically significant differences ($p > 0.05$) for sex distribution.

Table 2. Sex differences for dimensions of the matrix, bed and EDL tendon of the second toes ($n = 50$).

Outcome Measurements (mm)	Men ($n = 29$) Mean ± SD (Range)	Women ($n = 21$) Mean ± SD (Range)	p-Value
1. Nail matrix width at the proximal nail part (mm)	9.02 ± 1.09 (8.00–11.30)	6.77 ± 1.02 (5.50–8.20)	<0.001 †
2. Nail matrix width at the distal nail part (mm)	8.70 ± 1.90 (5.90–12.00)	8.92 ± 1.85 (5.80–11.80)	0.683 *
3. Length of matrix (mm)	2.39 ± 0.31 (1.80–2.80)	2.34 ± 0.28 (1.60–2.80)	0.535 *
4. Length of bed (mm)	5.67 ± 1.48 (3.90–8.70)	5.69 ± 1.31 (3.30–8.50)	0.555 †
5 Tendon width (mm)	2.41 ± 0.64 (1.50–3.50)	2.60 ± 0.68 (1.60–4.00)	0.215 †

Abbreviations: EDL, extensor digitorum longus; mm, millimeters. * Student t test for independent samples was used. † Mann–Whitney U test was used. For all analyses, statistically significant differences were set at $p < 0.05$.

Regarding lower limb side comparisons presented in Table 3, there were no statistically significant differences ($p > 0.05$) between left and right sides for any outcome measurement.

Table 3. Lower limb side differences for dimensions of the matrix, bed and EDL tendon of the second toes ($n = 50$).

Outcome Measurements (mm)	Left ($n = 26$) Mean ± SD (Range)	Right ($n = 24$) Mean ± SD (Range)	p-Value
1. Nail matrix width at the proximal nail part (mm)	8.21 ± 1.60 (5.70–11.30)	7.92 ± 1.48 (5.50–10.90)	0.785 †
2. Nail matrix width at the distal nail part (mm)	8.58 ± 1.70 (5.90–11.80)	9.03 ± 2.04 (5.80–12.00)	0.393 *
3. Length of matrix (mm)	2.42 ± 0.26 (1.90–2.80)	2.31 ± 0.32 (1.60–2.80)	0.194 *
4. Length of bed (mm)	5.87 ± 1.49 (3.30–8.50)	5.47 ± 1.29 (3.90–8.70)	0.398 †
5 Tendon width (mm)	2.60 ± 0.64 (1.60–4.00)	2.38 ± 0.67 (1.50–3.50)	0.398 †

Abbreviations: EDL, extensor digitorum longus; mm, millimeters. * Student t test for independent samples was used. † Mann–Whitney U test was used. For all analyses, statistically significant differences were set at $p < 0.05$.

This study analyzed the functional link of the nail to the second toe distal phalanx and various distal interphalangeal joint structures, such as EDL tendon fibers and collateral ligaments. EDL tendons continued from their bony insertions overlapping the distal phalanges, and the collateral ligaments formed an integrated network on both joint sides, thereby helping to anchor nail margins.

According to the findings obtained in our study, we have verified, through dissection and histologic studies, that EDL tendons ended at the dorsal area of the second toe distal phalanx and were attached to the complete phalanx, ending at the nail (Figure 2).

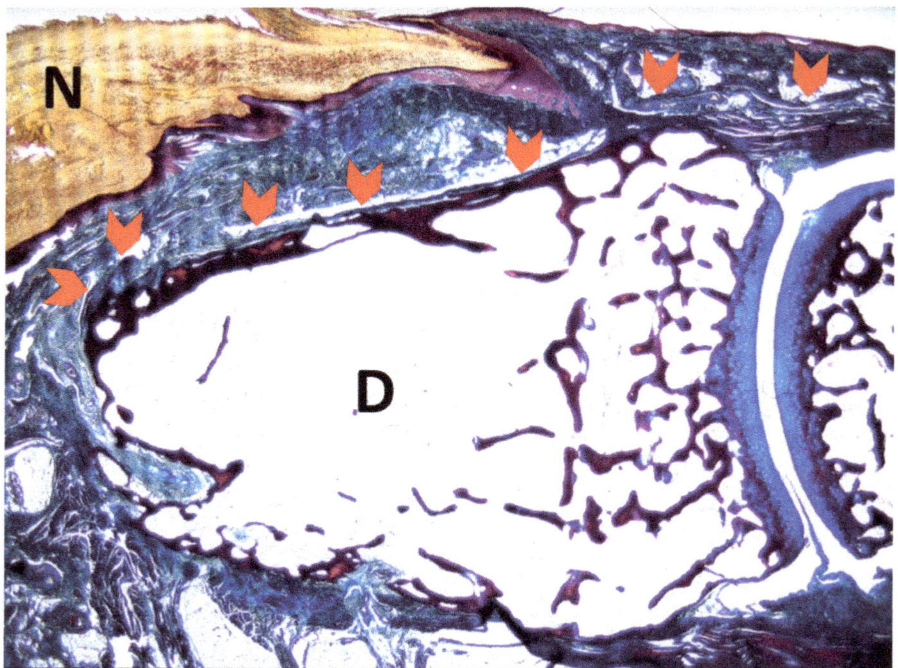

Figure 2. Sagittal section of the second toe at the midpoint of the nail matrix. The superficial fibers of the Extensor Digitorum Longus (EDL) tendon bundles extend to the dorsal aspect of the distal phalanx of the second toe (red arrowheads). N—nail plate; D—distal phalanx. (5× magnification). Tetrachrome VOF-III GS stain (light green SF/or fast green FCF, methyl blue, Orange G, and acid fuchsin). Scale bar = 5× magnification

Indeed, there was a close relationship between EDL tendons and the matrix. EDL superficial fibers were attached to the plantar matrix base underlying the distal phalanx (DF) bone and crossing the EDL deep fibers. EDL superficial fibers were expanded to the eponychium (Figure 3), which ran distally along the complete dorsal region of the distal phalanx.

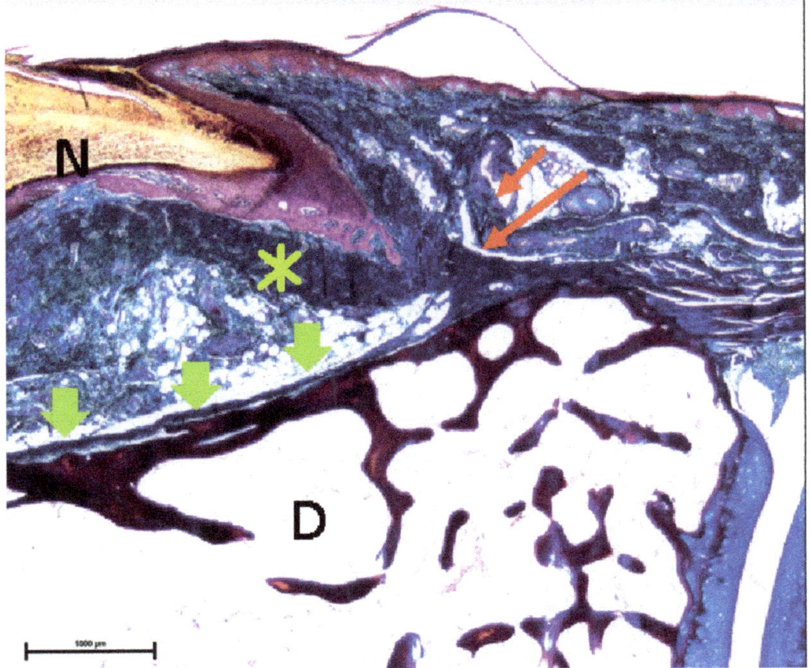

Figure 3. High-power magnification detail of the superficial fibers of the Extensor Digitorum Longus (EDL) bundles (red arrows) attached at the base of ventral matrix (asterisk) to the underlying distal phalanx (D) bone, crossing the deep fibers of the EDL tendon bundles (green arrowheads). Superficial fibers of the EDL tendon expansion to the eponychium (red arrow), which run distally along the complete dorsal aspect of the distal phalanx until its end (20× magnification). Tetrachrome VOF-III GS stain (light green SF/or fast green FCF, methyl blue, Orange G, and acid). N—nail plate; D—distal phalanx. Scale bar = 20× magnification

4. Discussion

Nails may be considered as specialized parts of the skin, such as an "epidermal appendage". Indeed, nails seemed to be functionally integrated with the musculoskeletal system. The second toe may play a key role in both morphological and biomechanical foot functions. At standing position, second toes showed more pressure than the fifth metatarsal heads and heels [2]. In addition, great toes produced a pressure about twice of the total pressure of the other four toes. Nevertheless, second toes supported the largest pressure after the hallux. During walking activities, as great toes and second toes were passively dorsiflexed, foot longitudinal arches were raised, the rearfoot was supinated, legs were externally rotated, and plantar aponeurosis was tensed [18]. Great toes and second toe disorders could modify subjects' static and dynamic balance, as well as gait, stance, and the entire erect bipedal locomotion process.

Toenails may play a key functional role regarding proprioceptive inputs, protective features, and control of toe pulps. Toenails seem to be related to the matrix of the distal phalanx periosteum. Matrix damage may lead to nail plate alterations [19].

The matrix location with respect to the EDL tendon may present clinical implications. Several studies have detailed the close relationship between the EDL tendon and the nail matrix, although prior studies have not delimited the exact anatomic distance in the foot [5,6,8–11,13,20]. These studies have been carried out in the hand and thus there are a lack of landmark studies in the foot region.

Authors have not found a similar study that quantifies the distances between the nail matrix and the EDL tendon insertion in the second toes.

A study of the matrix surface anatomy of the finger nail carried out by Reardon et al. [10]. delimited the distance from the nail bed proximal edge with respect to the distal interphalangeal joint. While these investigators showed that the extensor tendon insertions were always more proximal to the most proximal part of the matrix, no attempts were performed to quantify this distance. Regarding ingrown nails, a similar research work examined the matrix extent of toenails and reported that the matrix was extended to the extensor tendon insertion, although the distance was not also quantified [11]. Thus, our anatomic and histological study demonstrated the proximal edge of the matrix in the human cadavers second toes and the distal EDL tendon bony insertions.

A study limitation was that the nail plate proximal origin was delimited as the nail matrix proximal limit in order to determine its distal length, and thus the matrix was not evaluated more proximally than the nail plate. This could be a possible limitation, as the nail matrix may be extending more proximally than the nail plate itself. According to the sagittal section of the second toe at the midpoint of the nail matrix (Figure 2; Figure 3), often the nail matrix may present lateral matrix horns which may be extended as far proximally as the 75% of the distance to the distal interphalangeal joint crease [21]. Further research studies are needed in order to establish more accurate measurements. Furthermore, the impact of formalin preservation and histological process could have influenced the tissue relationships and dimensions. Although formalin retained shape and size of organs and vessels, Genelyn and Imperial College London soft-preservation (ICL-SP) solution technique may faithfully mimic cadavers' joints compared to un-embalmed cadavers [22]. Thus, future studies should compare our findings using different embalming solutions on the human cadavers' tissues. Finally, a main limitation was that our study was only a descriptive study according to our aim, which was to detail the anatomic and histological features of the EDL insertion to the proximal nail matrix of the second toe. Therefore, future studies should compare differences among outcome measurements such as the nail matrix width at the proximal nail part, the nail matrix width at the distal nail part, the length of the matrix, the length of the bed, and the tendon width. Several debilitating dermatoses may affect the nail unit, including papulosquamous disease and depigmented skin conditions.

5. Conclusions

In conclusion, the nail matrix and bed proximal edges of human cadavers' second toes were positioned dorsally and overlapped the distal EDL tendon until its distal bony insertion, ending at the nail in all human cadavers' feet. EDL tendons were not directly connected to the nail matrix due to the nail bed corium and nailbed mesenchyme being located between the matrix or nailbed epithelium and EDL. These anatomic and histological features should be used as reference landmarks during digital surgery and invasive procedures.

Author Contributions: Conceptualization, P.P.-L., M.E.L.-I. and R.B.-d.-B.-V.; Data curation, M.E.L.-I.; Formal analysis, M.E.L.-I., R.B.-d.-B.-V. and C.C.-L.; Investigation, P.P.-L., R.B.-d.-B.-V., J.M.-G. and D.L.-L.; Methodology, P.P.-L., M.E.L.-I., R.B.-d.-B.-V., D.R.-S., C.C.-L., J.M.-G. and D.L.-L.; Resources, J.M.-G.; Supervision, M.E.L.-I., R.B.-d.-B.-V. and J.M.-G.; Writing—original draft, P.P.-L., D.R.-S. and C.C.-L.; Writing—review & editing, M.E.L.-I., R.B.-d.-B.-V., D.R.-S., C.C.-L., J.M.-G. and D.L.-L. All authors have read and agreed to the published version of the manuscript.

Funding: This research received no external funding.

Conflicts of Interest: The authors declare no conflict of interest

Data Availability Statement: Data will be available upon request to the corresponding author.

References

1. Yamamoto, T.; Yoshimatsu, H.; Kikuchi, K.; Taji, M.; Uchida, G.; Koshima, I. Use of non-enhanced angiography to assist the second toetip flap transfer for reconstruction of the fingertip defect. *Microsurgery* **2014**, *34*, 481–483. [CrossRef] [PubMed]
2. Guy, R.J. The etiologies and mechanisms of nail bed injuries. *Hand Clin.* **1990**, *6*, 9–19. [PubMed]

3. De Berker, D.; Mawhinney, B.; Sviland, L. Quantification of regional matrix nail production. *Br. J. Dermatol.* **1996**, *134*, 1083–1086. [CrossRef] [PubMed]
4. Zook, E.G. Anatomy and physiology of the perionychium. *Clin. Anat.* **2003**, *16*, 1–8. [CrossRef]
5. Zook, E.G. Complications of the perionychium. *Hand Clin.* **1986**, *2*, 407–427.
6. Zook, E.G.; Van Beek, A.L.; Russell, R.C.; Beatty, M.E. Anatomy and physiology of the perionychium: A review of the literature and anatomic study. *J. Hand Surg. Am.* **1980**, *5*, 528–536. [CrossRef]
7. Ditre, C.M.; Howe, N.R. Surgical anatomy of the nail unit. *J. Dermatol. Surg. Oncol.* **1992**, *18*, 665–671. [CrossRef]
8. Guéro, S.; Guichard, S.; Fraitag, S.R. Ligamentary structure of the base of the nail. *Surg. Radiol. Anat.* **1994**, *16*, 47–52. [CrossRef]
9. Shrewsbury, M.; Johnson, R.K. The fascia of the distal phalanx. *J. Bone Joint Surg. Am.* **1975**, *57*, 784–788. [CrossRef]
10. Reardon, C.M.; McArthur, P.A.; Survana, S.K.; Brotherston, T.M. The surface anatomy of the germinal matrix of the nail bed in the finger. *J. Hand Surg. Br.* **1999**, *24*, 531–533. [CrossRef]
11. Hyder, N. Ingrowing toe nails: The extent of the germinal matrix. *J. Bone Joint Surg. Br.* **1994**, *76*, 501–502. [CrossRef] [PubMed]
12. Becerro de Bengoa Vallejo, R.; Losa Iglesias, M.E.; Jules, K.T. Tendon Insertion at the Base of the Proximal Phalanx of the Hallux: Surgical Implications. *J. Foot Ankle Surg.* **2012**, *51*, 729–733. [CrossRef] [PubMed]
13. Palomo-López, P.; Becerro-de-Bengoa-Vallejo, R.; López-López, D.; Calvo-Lobo, C.; Herrera-Lara, M.; Murillo-González, J.A.; Losa-Iglesias, M.E. Anatomic Association of the Proximal Fingernail Matrix to the Extensor Pollicis Longus Tendon: A Morphological and Histological Study. *J. Clin. Med.* **2018**, *7*, 465. [CrossRef] [PubMed]
14. Palomo Lõpez, P.; Becerro De Bengoa Vallejo, R.; Lõpez Lõpez, D.; Prados Frutos, J.C.; Alfonso Murillo González, J.; Losa Iglesias, M.E. Anatomic relationship of the proximal nail matrix to the extensor hallucis longus tendon insertion. *J. Eur. Acad. Dermatology Venereol.* **2015**, *29*, 1967–1971. [CrossRef] [PubMed]
15. von Elm, E.; Altman, D.G.; Egger, M.; Pocock, S.J.; Gøtzsche, P.C.; Vandenbroucke, J.P. The Strengthening the Reporting of Observational Studies in Epidemiology (STROBE) statement: Guidelines for reporting observational studies. *J. Clin. Epidemiol.* **2008**, *61*. [CrossRef]
16. Brewster, N.T.; Howie, C.R. Excision of the germinal matrix: A modification. *J. R. Coll. Surg. Edinb.* **1995**, *40*, 59.
17. Austin, R.T. A method of excision of the germinal matrix. *Proc. R. Soc. Med.* **1970**, *63*, 757–758. [CrossRef]
18. Tanaka, T.; Hashimoto, N.; Nakata, M.; Ito, T.; Ino, S.; Ifukube, T. Analysis of toe pressures under the foot while dynamic standing on one foot in healthy subjects. *J. Orthop. Sports Phys. Ther.* **1996**, *23*, 188–193. [CrossRef]
19. HICKS, J.H. The mechanics of the foot. II. The plantar aponeurosis and the arch. *J. Anat.* **1954**, *88*, 25–30.
20. Keyser, J.J.; Littler, J.W.; Eaton, R.G. Surgical treatment of infections and lesions of the perionychium. *Hand Clin.* **1990**, *6*, 137–153.
21. Jellinek, N.J.; Rubin, A.I. Lateral Longitudinal Excision of the Nail Unit. *Dermatologic Surg.* **2011**, *37*, 1781–1785. [CrossRef] [PubMed]
22. Balta, J.Y.; Twomey, M.; Moloney, F.; Duggan, O.; Murphy, K.P.; O'Connor, O.J.; Cronin, M.; Cryan, J.F.; Maher, M.M.; O'Mahony, S.M. A comparison of embalming fluids on the structures and properties of tissue in human cadavers. *J. Vet. Med. Ser. C Anat. Histol. Embryol.* **2019**, *48*, 64–73. [CrossRef] [PubMed]

MDPI
St. Alban-Anlage 66
4052 Basel
Switzerland
Tel. +41 61 683 77 34
Fax +41 61 302 89 18
www.mdpi.com

Diagnostics Editorial Office
E-mail: diagnostics@mdpi.com
www.mdpi.com/journal/diagnostics

www.ingramcontent.com/pod-product-compliance
Lightning Source LLC
LaVergne TN
LVHW070922170726
843515LV00005B/848